The Entangled God

Routledge Studies in Religion

The Entangled God

Divine Relationality and Quantum Physics

Kirk Wegter-McNelly

Routledge
Taylor & Francis Group
NEW YORK   LONDON

First published 2011
by Routledge
Third Avenue, New York, NY 10017

Simultaneously published in the UK
by Routledge
2 Park Square, Milton Park, Abingdon, Oxon OX14 4RN

*Routledge is an imprint of the Taylor & Francis Group,
an informa business*

Typeset in Sabon by IBT Global.
Printed and bound in the United States of America on acid-free paper by
IBT Global.

Library of Congress Cataloging-in-Publication Data
Wegter-McNelly, Kirk, 1967–
    The entangled God : divine relationality and quantum physics / by Kirk
Wegter-McNelly.
       p. cm. — (Routledge studies in religion ; 15)
    Includes bibliographical references and index.
    1. God (Christianity)   2. Quantum theory—Religious aspects—
Christianity.   I. Title.
    BT103.W44 2011
    261.5'5—dc22
    2011006331

ISBN13: 978-0-415-77558-8 (hbk)
ISBN13: 978-0-203-80592-3 (ebk)

For Jennifer,
who has taught me much about
the relational intricacies of
partnering and parenting

Contents

Figures

Acknowledgments

The origins of this book date to the early 1990s, when I first discovered the scholarly world of "religion and science." Having decided to explore the possibility of making a career out of my newfound interest, I enrolled at Princeton Theological Seminary (PTS) to study with Wentzel van Huyssteen, who had joined the faculty only a semester earlier as the first James I. McCord Professor of Theology and Science. My time at PTS working with Wentzel was highly formative, both in terms of my broader theological outlook and in terms of my growing awareness of religion-and-science-related issues. Wentzel was a gracious advisor and mentor, and I remain grateful for his theological and professional guidance over the years. Others who made studying at PTS a rewarding experience include Janet Abel, Marne Arthaud-Day, David Janzen, Michael Gross, Bill Locke, Terri Luper, Mark McKinney, Jill Russell, Tricia Sheffield, Joyce Walker, and Michael Wilson.

My first attempts to think about the theological significance of entanglement took place in Berkeley, California, where I was doctoral student at the Graduate Theological Union (GTU) from 1995 to 2003. Among the many people whose friendship sustained me during my time there, I would like to thank Susan Ashbourne, Anne Badé, Jaime Balboa, Gaymon Bennett, Diane Bowers, Greg Cootsona, Scott Dickerman, Chuck Eyler, Chuck Goodman, Laine Harrington, Janet Holmes, Dennis Keller, Kevin Koczela, Kathleen Kook, Bob Lasalle-Klein, Lonna Lee, Cheryl Maddox, Marty Maddox, Maureen Malony, Marion Mullin, and the other members of the Ohlone Chambersingers, Moses Penumaka, Brad Peterson, Jane Redmont, Charlotte Russell, Nancy Saltsman, Fred Sanders, Susan Sanders, Dotty Timourian, and Hector Timourian. The hub of daily life in Berkeley, for me at least, was the Center for Theology and the Natural Sciences (CTNS) at 2452 Virginia Street, where I had the pleasure of working with Philip Clayton, Chris Doran, Nathan Hallanger, Peter Hess, Bonnie Johnston, Stan Lanier, Joshua Moritz, Melissa Moritz, Norrie Palmer, Ted Peters, Richard Randolph, Mark Richardson, Robert John Russell, Jim Schaal, Cathy Thompson, Lou Ann Trost, Holly Vande Wall, Marc Wallman, and Carl York, as well as many other visitors who took advantage of one or more of the center's many conferences and programs, including Ian Barbour, Paul Chung, Paul Davies, Christoph Lameter, George Murphy, John Polkinghorne, Drew Porter, Abner Shimony, Christopher Southgate, William Stoeger, Margaret Wertheim, Adrian Wyard, and Joseph Zycinski.

During my time in Berkeley I had the honor of attending several conferences on divine action sponsored jointly by CTNS and the Vatican Observatory and held at Castel Gandolfo in Italy. George Coyne, the director of the Observatory at the time, was always a most gracious host. My own ideas about the entangled mode of God's presence in creation are deeply indebted to the noninterventionist theological explorations of the participants at these conferences, especially those of Philip Clayton, George Ellis, Nancey Murphy, Arthur Peacocke, John Polkinghorne, Robert Russell, William Stoeger, and Thomas Tracy. Above all, my sincerest thanks go to Russell, the founder and director of CTNS and the first holder of the Ian G. Barbour Chair in Theology and Science, who was my doctoral advisor, along the way became my mentor, and continues to be a generous source of professional and personal support despite the fact that we now live on opposite coasts. Bob's uncommon expertise across the disciplines of theology and physics drew me to the GTU, and his remarkable creativity, vision, kindness, intellect, and spirit kept me there. He is in no small part responsible for creating and nurturing the field in which I now teach and write. I cannot imagine having had a better or more gracious mentor. Many of the ideas presented in this book were percolating in Bob's mind long before I knew anything about the problem of divine action or the significance of quantum entanglement. I am most grateful for his continued support, especially for his thorough and careful reading of the penultimate draft of this book. I also want to thank the other members of my doctoral dissertation committee, who were helpfully critical and genuinely encouraging: Raymond Chiao, William Stoeger, and Ted Peters. Still others have been kind enough to engage me in scholarly dialogue since leaving Berkeley, including Graham Buxton, Reinhard Erdmann, Don Howard, John Roche, Ernie Simmons, and Taede Smedes.

After completing my doctorate in 2003, I transitioned into full-time teaching at Manhattan College (MC) in the Bronx, New York. My time at MC was a real pleasure, thanks to students, colleagues, and friends who gave me such a hospitable welcome when I arrived, and who (very) graciously wished me well when I departed for Boston University. At other points in my life I have had to work for many years to build up circles of friendship; at MC I was instantly welcomed into an intellectually vibrant group of colleagues who knew how to think, argue, teach, and . . . how to have a good time doing it all. I regret no longer having the benefit of their daily companionship and wit. I especially want to thank Martha Ackelsberg, Jorge Afanador, Jennifer Butler, Ashley Cross, Stephen Kaplan, John Keber, Jeff Horn, Judith Plaskow, Rocco Marinaccio, Michele Saracino, and Glenn Zuber.

In the summer of 2004 my wife, Jennifer, and I moved to Boston. We arrived only three weeks, it turned out, before the birth of our son, Keller. A number of people lightened the burden of getting to know Boston and offered their friendship and support at crucial moments, including Faythe Beauchemin, Ian Bouchard, Sarah Bouchard, Eden Brenner, Al Circeo, Cavan Concannon, Dave Courage, Sam Courage, Jessica Haffner, Rae Huang, Megan Humphreys-Loving, Pippa Mpunzwana, Costas Rodopoulos, Arnold Rots, Betty Rots, Burns Stanfield, Lorraine Stanfield and Suzanne Wildman. When I joined Boston University's School of Theology (BU-STH), both the school and the university were in a time of transition. I was excited and honored to find myself part of such a thoughtful, productive, and committed group of students, administrators, and scholars. I have since come to find out, thanks to the support and encouragement of many in the school, that I have a penchant for institution-building. I have also been told that there is something deeply Confucian about my soul. Among so many students, staff, and colleagues, I am especially grateful for the collaborative spirit of Joas Adiprasetya, Alisa Bokulich, Peter Bokulich, Christopher Brown, Jackie Ammerman, Nancy Ammerman, Choi Hee An, Dale Andrews, Nat Barrett, Alejandro Botta, Wesley Dalton, Inus Daneel, Katheryn Pfisterer Darr, Brian McGrath Davis, Carl Daw Jr., Christopher Evans, Norm Faramelli, Walter Fluker, Susan Forshey, Sarah Fredericks, Julian Gotobed, Garth Green, John Hart, Susan Hassinger, Robert Hill, Megan Hornbeek, James Jackson, Gregg Jaeger, Sam Johnson, Maggie Keelan, Anastasia Kidd, Chad Kidd, Kim Han-Kyung, Jennifer Knust, Christopher Lehrich, Derek Michaud, Lourey Middlecamp, Robert Neville, Rick Peters, Rodney Petersen, Valentina Pride, Thomas Porter, Jr., Shelly Rambo, Dana Robert, Court Randall, Chris Schlauch, Andrew Shenton, Anjulet Tucker, Karen Westerfield Tucker, James Walters, and Claire Wolfteich.

There are a number of people at BU-STH who deserve a special thank-you. In particular, I owe a debt of gratitude to Deans John Berthrong, Ray Hart, Mary Elizabeth Moore, Imani-Sheila Newsome, and Bryan Stone, whose support I have relied upon heavily since arriving in Boston. I am especially thankful for the efforts of my senior colleague, Wesley Wildman, who played a crucial role in bringing me to BU-STH and who, in the midst of a very busy schedule, provided careful and timely comments on the first full draft of this book. I appreciate his unflagging yet patient efforts to help me succeed as a teacher and a scholar, as well as his entrepreneurial spirit in bringing back to life and then leading BU's master's and doctoral programs in religion and science. Wesley has an astonishing ability to see quickly and sympathetically multiple sides of an issue; I look forward to deepening our scholarly relationship in the years ahead. One additional group within the BU-STH community deserves special mention, namely, those who took part in the course I taught on this book in the spring of 2010 at Dean Moore's suggestion. As she predicted, teaching the book was an invaluable experience. I learned a great deal about it and myself from

Jeff Dodge, Eric Dorman, Victoria Gaskell, Paul Morris, Nathan Strunk, Ben Thompson, Sharon Thornton, and Larry Whitney. Nathan, who was my teaching assistant for the course, did yeoman's work in transcribing a digital recording of every class session. This valuable resource became its own kind of manuscript and numbered into the hundreds of pages. Nathan also deserves credit for his part in designing and implementing the book's index. The book was shaped at various other stages by my other research assistants: Chi Sang Woo, Wesley Dalton, and Josh Reeves. Sang Woo, in particular, was a highly conscientious and skillful assistant. He, Nathan Strunk, and Connor Wood read and commented on the penultimate draft of the book, and Wes offered helpful comments on the first five chapters. One other BU-STH student, Ian Mevorach, read and commented on the penultimate draft of the theological chapters.

As I was working on the entanglement project I became heavily involved in an interdisciplinary research project on the theological significance of gravitational-wave physics, thanks to a generous grant I received through CTNS's innovative grant program called the "Science and Transcendence Advanced Research Series." My work on that project helped me see much more clearly what I needed to do in this book and how I needed to do it. I am profoundly thankful for everything I have learned from my co-principal investigator, Raymond Chiao, and his team, which included among others Spencer DeSanto, Luis Martinez, and Stephen Minter. Ray has the patience of a teacher who truly loves to teach. I am indebted to him for generously sharing his vast knowledge and research skills with me in person, by email, and over the phone for the past four years.

Several names that might have appeared at various points in the preceding were absent because I wanted to acknowledge their significance to my life in a separate paragraph. I have been blessed with the gift of becoming close friends with a number of people, all of whom have celebrated the joys in my life and some of whom have sat with me in the midst of death. I remain humbled by their desire to know me as I really am, and I give thanks for the many ways in which their friendships enrich my life. Those whose relationships I cherish most—but don't do enough to nurture—include Jane Austin, Larry Gordon, Susan Hylen, Preston Klassen, Mary Lowe, Nancy Pineda-Madrid, Andrew Shahriari, Eric Sickler, Ted Smith, and Greg Zuschlag.

I am also fortunate to have an extended family that delights in my foibles and occasionally defers to my good judgment. I would not have made it this far personally or professionally without the support of Barbara Black, Bill Boulger, Sara Boulger, Page Boulger, Kathy Carlson, Richard Carlson, Marsha Huha, Arloa Hymans, Nelson Hymans, Terry Hymans, Sylvia Hymans, Jack McNelly, Lynn McNelly, Nancy Slobodzian, Terry Slobodzian, Maggie Slobodzian, Jessica Slobodzian, Emma Stenner, Eric Stenner, Weg Stenner, Lena Tom, Phil Tom, Diane Wegter, Todd Wegter, and the Northwest-Iowa Wegter clan. Within this larger circle, I have often sought out the counsel and harmonies of my sister, Heather Stenner, and my aunt, Sandy Brooks. Heather keeps alive many of the good things that were important to our parents, Charlene and Robert. Sandy has stepped into the role of parent, a role which she has taken deeply to heart—and for her willingness to do so I will always be grateful. Finally, I want to thank my partner, Jennifer, and my son, Keller, for constantly renewing and expanding my life in myriad ways. They inspire me in their work and in their play to think more deeply about how to live faithfully and in right relationship. I hope I can continue to grow into the person they deserve to have as partner and father. Their patient encouragements and watchful eyes have put me back on track more than once.

A final word of thanks to Lesley Riddle, Laura Stearns, and Stacy Noto at Routledge Press, as well as Eleanor Chan at IBT Global, for their expert guidance during the process of publication.

1 Setting the Stage

> Then God said, "Let us make humankind in our image, according to
> our likeness."
>
> Genesis 1:26a (New Revised Standard Version [NRSV])

Since the time of the early church, Christians have wrestled with the idea
that creation's "otherness" vis-à-vis God does not contradict but rather
reflects the fullness of God's infinite being. How can something "other"
exist in relation to the infinite? According to the trinitarian thinkers of the
early church, otherness is basic to God. The being of the One whom Jesus
called *Abba*, said the church's first theologians, is "Being-in-Relation"—an
idea that achieved technical expression in the notion of the "perichoretic
mutuality" of the trinitarian persons within the Godhead. Subsequent generations came to associate their belief in God's faithfulness across human
history with the fundamentally faithful character of God's primal decision
to grant existence to creation as "other." Within Christian thought God's
identity as creator has come to be regarded as an outwardly, otherly, and
faithfully fecund expression of the inner-eternal life of God, within which
otherness already is and always has been.

Each new generation must take up the challenge of reflecting on what it
means to say that God is relational, that God's interaction with the world—
even the very act of bringing it into existence—faithfully reflects who God
is. The relational view of divinity that grew out of the church's early struggles to understand what it means to say that Jesus reveals God has reap-
peared with new emphases and accents in our own day. This vision of God
can be found in a variety of contexts, most prominently in the widespread
desire to norm the constructive and destructive dimensions of human relationships (both among humans and with the planet) against a larger vision
of sustaining mutuality.

This book aims to contribute to ongoing theological discussions about
divine relationality by introducing a theological audience to one of the most
significant scientific discoveries of the twentieth century—the detection of
correlated or "entangled" behavior in the physical world. Variously interpreted as evidence for the existence of faster-than-light communication,
instantaneous physical connection between objects persisting across arbitrarily large distances, the coalescence of distinct physical objects into a
larger, nonseparable whole—and in yet other ways, as we shall see—the
discovery of entangled relations among physical objects calls into question

our everyday intuitions about what it means to exist in the world as distinct and separate "individuals."

Those who are unfamiliar with the term "entanglement" or who associate physics with the low point of their high school years may approach this book with some trepidation. It would be a mistake, however, to think that the mathematical and philosophical portions of the book are intended for the scientific insider. The story told in the following pages is for anyone who thinks that serious interdisciplinary engagement with the sciences is not beyond what can or should be expected of today's student of theology. It aims to make such engagement possible for those with minimal exposure to scientific and mathematical reasoning. Granted, this book may not occasion the kind of experience one typically associates with reading a theological text, but that is because the interdisciplinarity promoted herein implicates not only the act of writing but also the act of *reading*. This is not just a book with interdisciplinary origins, but one meant to facilitate an interdisciplinary *experience*. Living into the deepest questions of existence requires determined engagement with multiple ways of knowing (Belenky et al. 1986).

The story of entanglement is a story worth knowing in its own right, regardless of one's own theological proclivities. It is the tale of several generations of physicists reluctantly coming to grips with a peculiar type of relationality among physical objects. After two, say, electrons have become "entangled" with one another, they unfailingly modulate their own subsequent behaviors to account for whatever their companion does, no matter how widely separated they might come to be. What's more, the entangled nature of their relationship can be seen only "globally" in the joint behavior of both electrons, not "locally" in the individual behavior of either one. In brief, the story of entanglement takes one beyond the classical, Newtonian world of billiard balls bumping around blindly and predictably into a quantum world of subtle relationality and effective unpredictability. Entanglement is well on its way to being regarded as one of the most important discoveries ever made within the history of physics—on par with those of Copernicus, Galileo, Newton, and Einstein. The empirical confirmation of entanglement, which Alain Aspect has dubbed "the second quantum revolution" (see his preface to Scarani 2006, xii), carries immense potential to stir new constructive theological work on the nature of divine and creaturely relationality.

The overarching purpose of this book is to probe the history and interpretations of entanglement for the sake of developing the concept as a theological metaphor for God's relational being and the relation between God and creation. As a part of that larger purpose, the final chapter considers the traditional theological question of how God acts. Against the ideas that God creates the world out of undifferentiated and absolute nothingness (*creatio ex nihilo*), on the one hand, and that the world flows naturally and without any intentionality from God's plenitudinous being (*emanatio ex deo*), on the other, I argue that a trinitarian account of creation should begin with the idea that God creates the world out of the divine relationship (*creatio ex relatione*). On this view, the world exists not by the act of some generic God, but rather by the act of the trinitarian, entangled God who "relates" or "entangles" into existence a world characterized by its own modes of relationality. I then deploy the metaphor of entanglement to characterize God's ongoing presence in the world, especially in light of the pervasiveness and persistence of evil and suffering. A God who is entangled with the world does not overwhelm but allows each part of creation to be what it is, all along loving it into the fullness of what it might yet become.

The shape of the trinitarian argument laid out in the preceding reflects the overall epistemic posture of the project; namely, it begins from an explicitly trinitarian stance in order to consider afresh the relational nature of the Christian God in conversation with contemporary scientific and philosophical reflection (cf. Polkinghorne 2010b). Such an approach has aptly been called a "theology of nature" (Barbour 1997, 98–105; see also Wegter-McNelly 2000) and should not be mistaken for an *au courant* "natural theology" bent on leveraging the latest scientific discovery (entanglement) to warrant a particular theological claim about the character of ultimate reality (trinitarian relationality). "Natural theology" and "theology of nature" draw their epistemic inferences in opposite directions, with the latter rooting theological conviction primarily in communal life and discernment rather than in empirical observation.

What is broadly at stake in these pages is the standard to which Christians ought to hold themselves when calling God "creator." If we fail to use this term in ways that connect to our awareness and understanding of the world as it is revealed through the sciences, we risk invoking a god who created some imaginary world instead of the God who created ours. Sallie McFague has said that theology "is not about 'God and the world,' but about God and a particular world, some concrete interpretation *of* the world" (2000, 71; italics in original). We need to be clear about what particular world we have in mind when we speak of God as its creator. When we call God "creator," we speak "creation" in the same breath, and vice versa. Thinking theologically about entanglement opens an intriguing possibility: God's own relationality becomes the source of *all* creaturely relationality—not just our own, nor that of all life, but also that which characterizes the physical world in general.

It is worth pondering at the outset whether Christians who live long after the revolutions inaugurated by Newton, Darwin, and Einstein still experience the physical world as creation. To put the matter in the form of a question: Do Christians today find any theological meaning in the intricacies of the physical world? When the needs of "the whole world" are lifted up in worship, how broad is the intended scope of reference? The Earth? Only humanity and other forms of terrestrial life? What does it mean to speak of the "whole" world? According to the Gospel of John, all of creation has its origin in Christ, the *Logos* of God: "All things came into being through him, and without him not one thing came into being" (John 1:3, NRSV). Within the Pauline tradition, too, the world coheres in the one through whom it was made: "all things have been created through [Christ] and for him. He himself is before all things, and in him all things hold together" (Col. 1:16b–17, NRSV). The biblical writers understood their world to be something more than an agglomeration of disparate realities. They understood it to be a unified whole with a single origin and purpose, not through its own power but through the power and will of its creator. What will it take to bring the physical world back into our theological field of vision (cf. Tracy 1981, 215)? How can we "re-create" the physical world in our own theological imaginations so as to reintegrate it into the stories we tell about God as its creator? The rise of modern science in the early modern period did much to dissolve medieval theology's clear perception of creation's integral wholeness, replacing this worldview with a reductionistic vision that—it must be said—allowed scientists to make enormous progress in understanding how the world works. But in recent decades the science of quantum mechanics has begun to bump into the limits of reductionism. From a theological point of view, the discovery of the physical phenomenon of quantum entanglement comes as a glimmer of relational wholeness that might allow us to speak of the world around us as something much more than a mere agglomeration.

1   THE SCRIPT

The narrative structure of this book proceeds from theological beginnings to theological endings by way of intermediate chapters on the history, science, and philosophy of entanglement. The opening theological moment begins in Chapter 2 with a consideration of the recent convergence in theological writing around the topic of "divine relationality." Turn the pages of almost any contemporary Christian theological text on God's nature and activity in the world—whether by a process, feminist, postliberal, liberation, trinitarian, ecological, evangelical, political, or Eastern Orthodox theologian (and the list could go on)—and one finds "relationality" both identified as the lasting inheritance of the trinitarian tradition for contemporary theology and invoked as the key category for understanding God's nature, purposes, and presence in the world. The purpose of this chapter is not to argue for a new view of God but to demonstrate that "divine relationality" has indeed become a major theme among Christian theologians over the past fifty years. (An appendix at the end of the book provides a list of recent monographs treating the subject at length.) Those who have engaged the sciences on the topic of relationality typically work with images and metaphors from classical (i.e., pre-quantum) science, whether physics or biology, unwittingly drawing upon a scientific paradigm unfriendly to the idea that being is constituted (whether partly or wholly) by relations. From the classical perspective, relations have a strictly secondary ontological status and thus a limited role in how the world works; the possibility of being a cause lies primarily with the particulars of the world, i.e., with individuals instead of the relations among them. Individuals may interact in highly nonlinear ways that give rise to a kind of holistic behavior, but as objects in the world individuals remain individuals. The classical sciences thus offer a limited range of possibilities for thinking more holistically about relationality.

Chapter 3 explores an important empirical consequence of the *classical* (i.e., pre-quantum) perspective. At the heart of the chapter are three hallmark principles of classical physics: definiteness, separability, and locality. The chapter begins with a brief and thoroughly classical account of "photons," i.e., particles of light. This opening section serves as a touchstone for subsequent discussions. The three hallmarks of definiteness, separability, and locality are especially helpful for understanding the famous objection leveled against quantum theory in 1935 by Einstein and his two colleagues, Boris Podolsky and Nathan Rosen (hereafter, EPR). They argued that if quantum theory is a complete description of physical processes, then something like entanglement must obtain (EPR 1935). Thinking that no sane person could possibly accept such a problematic conclusion—Einstein disparagingly called entanglement "spooky action at a distance [*spukhafte Fernwirkungen*]" (Born 1971, 158)—they reasoned that quantum theory provides only a partial description of physical processes. The climax of this part of the story came in the 1960s when another physicist, John Bell, recast EPR's objection to quantum theory into empirically testable form (1964). The fruit of Bell's labor, commonly called "Bell's theorem," requires a particular experimental outcome *if the world follows classical rules*. Sophisticated variations on Bell's theorem have since been constructed, but the basic force of the theorem in its simplest form can be apprehended without any advanced mathematics—only algebra and the smallest bit of trigonometry. Part of the interdisciplinary experience of reading this book involves reconstructing a simple version of Bell's theorem in just a few pages. The theorem is one of the greatest achievements of twentieth-century physics, and anyone with a basic education and a modicum of determination can understand it.

Whereas the ideas introduced in Chapter 3 all fit tidily within the world of classical physics, the ideas and experiments introduced in Chapter 4 do not. Enter: the strange world of quantum physics.[1] The results of the many Bell-type experiments that have now been performed in laboratories around the world clearly violate the outcome predicted by Bell's theorem. What's more, they agree precisely with the prediction given by quantum theory. The weight of evidence overwhelmingly confirms the existence of entanglement in the world—to the apparent demise of the "commonsense" view of classical physics epitomized in the three principles of definiteness, separability, and locality. But how can this be? In particular, how does quantum theory "see" the world such that it flouts Bell's prediction?

Chapter 4 takes up this important question. First we review some of the key experiments that established the existence of entanglement. Then we identify the key conceptual shift that distinguishes the quantum perspective from that of all classical physics: quantum theory's use of the principle of "superposition." With this important principle in mind, we reconsider photons from a quantum perspective and show how to predict their behavior using quantum theory. Finally, we examine how these ideas lead to the quantum prediction for an actual Bell-type experiment. The results obtained by physicists since the early 1980s overwhelmingly confirm the quantum prediction, leaving little room for doubt that quantum theory gets something profoundly right about the world in which we live. Whereas the task of this chapter is to show how physicists relate experimental outcomes to mathematical predictions, the task of the following chapter is to think carefully about what the predictive successes of quantum physics mean for our understanding of the physical world.

To this end, Chapter 5 introduces readers to several of the philosophical interpretations of entanglement that have been put forward in recent years. Some physicists and philosophers have argued that the mysterious nature of entanglement disappears when one adopts a particular interpretation of the larger theory. Others have argued that, in fact, no explanation is needed. The only thing in need of adjustment, they say, is our own general sense of the capabilities of scientific theories. If we need not squeeze the "way things are" out of our theories for them to be useful, then why think of quantum correlations as a problem? They just happen. Still others have tried to salvage the classical worldview by imagining minimally disruptive ways of modifying the classical framework so as to get around Bell's theorem. A fourth group, whose work I find especially interesting, argues that entanglement reveals the existence of a particular species of relational holism that I label "nonemergent, nonsupervenient holism." There is presently no definitive argument for or against any of these views, although future evidence may tip the scales in one direction or another. My own view of entanglement builds on the relational-holist idea that complex entangled systems have properties carried by the relationships among the parts but not by the parts themselves. I conclude the chapter by considering several aspects of entanglement from a relational-holist perspective.

In Chapter 6 I bring my own relational-holist interpretation of entanglement to bear on a trinitarian-relational view of God and the God–world relation. The chapter explores the theological ramifications of characterizing the Christian trinitarian God as "entangled," chiefly in terms of possible connections between the nature of God's being and (1) God's creative act of bringing the world into existence, (2) the character of the world, (3) the mode of God's presence in the world, and (4) the advantages of a plerotic (*plerosis* = "fullness") account of divine action over a kenotic one.

The image that emerges is one of a God whose plerotic, entangled presence underwrites the freedom of creation to be and become itself, even as this presence transformatively brings creation's behavior into accord with God's own activity. I conclude with a brief reflection on the future of "entanglement" as a theological metaphor.

The theological proposal developed across these chapters makes itself empirically vulnerable insofar as it appeals to the plausibility of the relational-holist interpretation of entanglement as the basis for its own view of God and God's relation to creation. If the relational-holist interpretation were to be shown unsatisfactory in some significant way with respect to further scientific or philosophical work, this would undermine the credibility of the theological proposal. Connections made across disciplines must be advanced in full awareness that future developments could once again require radical rethinking (cf. Russell 2001, 301–305; for the contrary view, see Peacocke 1993, 28; McGrath 2001, 45–50), but contemporary theology ought to welcome the liability that comes with searching for points of contact on the edge of what is known rather than at the center of what is familiar and established. A tradition—theological or otherwise—sustains itself as a *living* tradition only by continually placing the insights and wisdom it identifies with its past into conversation with the exigencies and opportunities of an ever-changing present. To engage the sciences with this type of awareness from a theological perspective is to realize that theological assertions must always be advanced with fallibilist—one might say "hypothetical"—intent. Like their scientific counterparts, theological theories must remain "open to revision in light of further research and discussion" (Clayton 1989, 166).

2   METAPHORICAL THEOLOGY—A NEW APPROACH

Metaphors and analogies, it has rightly been said, prove nothing. To suggest, for instance, that the portrayal of God's wrath in the Hebrew Bible can be understood as the anger of a loving mother toward a child who repeatedly engages in self-destructive behavior does not prove that God's anger is in fact like that of a loving mother, or even that God "gets angry" at all. Such an image, however, can release fresh insight into biblical accounts of divine wrath for the sake of a theological vision in which God is understood neither as spiteful nor as abusive but whose anger at sin reflects a loving desire for human flourishing. At their best, metaphors and analogies open up personal intuitions and viewpoints for public exploration and debate. They place our hunches out on the table for inspection, so to speak. Their power lies not in their ability to establish truth but in their capacity to (re)orient a community's efforts at making meaning out of its texts and traditions. "It's like this," we say, as we seek to convey our sense of how things really are. The power of a metaphor is not unlike the power embodied in an artist's attempt to gesture toward truth by pursuing an image or idea. But wherein does the plausibility of a theological metaphor lie? And if metaphors are incapable of warranting assertions, then what makes one metaphor "good" and another "bad"? Is it possible to speak of responsible and irresponsible theological uses of metaphor? In this section I address these questions by appealing to the notion of interdisciplinarity.

Sallie McFague, drawing chiefly on the work of Ian Barbour (1974) and Paul Ricoeur (1975), has argued that theological language—indeed, that all of human learning—has a metaphorical quality (McFague 1982, 34; for other contributions to the discussion, see Soskice 1988; Hesse 1988; Happel 1995, 1996). Metaphor, she notes, contains an important "is and is not" quality that must always be held in view. Forgetting a metaphor's "is not" leads to the literalization of truth, whereas forgetting its "is" leads to a mistaken renunciation of its truth-bearing potential. The task of living and speaking truth without veering toward either sterile extreme births the recognition that truth is an inseparable mixture of creation and discovery. This is the burden of all human attempts at knowledge.[2] Some theological metaphors prove to be particularly powerful images over time and thus come to function as touchstones of religious experience and sources of further theological development. Such powerful metaphors, or "models" as McFague calls them, are the stock-in-trade of religious reflection. The problem with theological models, as McFague sees it, is that they tend to exclude other models and insist upon the literalness of their own truth (1982, 23–24). She argues that theology cannot do without models and must therefore incorporate the metaphorical "is–is not" character of models into its methods in order to avoid the idol of literalization. More recently she has written, "Metaphorical language as theology's language is simply the acknowledgment that everything we say about God is partial" (2000, 30).[3] This book aims to increase the storehouse of models available to Christian theologians and communities, for only in plurality can models effectively counteract each other's limitations (their is-not-ness) and only in plurality can they speak effectively, through their partiality, to the inexhaustible depths of the divine life.

Any adequate theological metaphor must be conceptually consistent, capable of comprehensive application, similar or complementary to the major models of the theological tradition rather than opposed to them, and able to cope with anomalies (McFague 1982, 131–144). With regard to the last of these criteria, McFague advocates an empirical fit between models and the world, although she has "human" rather than "physical" data in mind. Following her lead, I propose an interdisciplinary way of doing metaphorical theology, one that takes the time to understand but also to explain the disciplines it engages. For theology's own sake, and for the sake of increasing general understanding about our world, theological scholarship, especially that which goes by the name of "theology and science," needs to become more deeply interdisciplinary in the sense described earlier—not merely in terms of the research process but also in terms of how its fruits are presented. This is a call for a certain kind of tethering—more attitudinal than inferential—both in research and in writing.

Theology must be allowed to speak in its own voice, with its own imagination, and through its own resources about God, humanity, and the world. The chief virtue of the metaphorical approach advocated here is that it allows theology to maintain its integrity *qua* traditional discourse in the midst of an interdisciplinary encounter. But an interdisciplinary metaphorical theology—one that dips into the wells of other disciplines for inspiration as it constructs its own images of God, humanity, and the cosmos—must be an eminently responsible theology. The integrity of metaphorical theology suffers, and there is little opportunity for deeper appreciation or appropriation, when it gleans what it wants from other disciplines without a keen sense of responsibility to engage and represent those disciplines as faithfully as possible within its own realm of discourse (cf. Van Huyssteen 1999, esp. 264ff.). The following statement by James Loder and James Neidhardt (1996, 283; italics in original) evinces this understanding of the theological task:

> [T]he inner nature of God cannot be pursued extrinsically, abstractly and deductively, as if we could deduce from an analysis of the physical universe the nature of God. However, this does not mean that theology must reject the sciences. Rather, it means that scientific investigation of the natural or the human order must be brought within the body of positive theology, and pursued in indissoluble unity with it. The sciences must become *natural* to the fundamental subject matter of theology; they will provide the inner material logic that arises in our inquiry and understanding of God.

Constructive theologians who wish to appropriate knowledge from other disciplines owe it to those disciplines, first, to understand and present them at a level that goes beyond basic acquaintance or general overview; second, to explain how the concepts and images being appropriated function within their native context; and third, to indicate how and why the concepts and images are being reshaped for theological purposes. The integrity of an interdisciplinary metaphorical theology comes not from adhering slavishly to original meanings but by providing a careful account of original meanings along with a clear justification of the ways in which those meanings are being sustained but also modified.

This book aims to fulfill these goals by introducing readers to the key developments in twentieth-century physics that led to the discovery of entanglement and by giving them the opportunity to encounter and understand the basic scientific details behind the narrative. To be sure, the larger purpose of the book is an unabashedly theological one: to think about God's relation to the world in light of what we know about the world through the natural sciences. But the integrity and vibrancy of any constructive theology, especially an interdisciplinary metaphorical theology, depends on its willingness to weave the stories of other fields and disciplines into its own narrative. Why? Because it is just too easy to ignore the complexities of the world on which we want to reflect theologically. One way to demonstrate that a theological point of view does justice to such complexities is to give readers the opportunity, within the theological text itself, to see and understand the world from the relevant extra-theological perspective. Taking the time to weave together narratives from multiple disciplines will leave less room for exclusively theological speech, but in this case less is potentially much more. A thickly interdisciplinary theology brings together the dialectical powers of consonance (McMullin 1981) and dissonance (Russell 1989) for the sake of establishing transversal connections that respect disciplinary differences (Van Huyssteen 1999).

Robert Russell has developed a vigorous proposal for interdisciplinary work in theology and science, which he calls "creative mutual interaction" (2002, 10–30). Russell identifies five distinct ways in which science appropriately influences theology: scientific theories (1) directly constrain theology, (2) serve directly as a constructive basis for theology, (3) indirectly affect theology through philosophical interpretations, (4) indirectly affect theology through a philosophy of nature, and (5) provide heuristic aid to theological thinking through conceptual or aesthetic inspiration. My own appropriation of quantum entanglement employs each of these approaches, emphasizing (3) and (5). The relational-holist interpretation of quantum entanglement does not yet constitute a robust philosophy of nature, but recent interpretive work has begun to lay the relevant groundwork.

What distinguishes Russell's methodological proposal from others is that he also identifies three ways in which theology can be understood to appropriately influence the sciences: theological theories can (1) warrant philosophical assumptions underlying scientific methodology, (2) provide heuristic insights for setting research agendas, and (3) offer testable implications that can motivate scientists to pursue particular research agendas. I would like to propose one addition to this list: (4) Theological writing can provide a detailed yet accessible account of the relevant scientific details for a nonspecialist audience. Whereas Russell's three paths of influence are all direct, mine is indirect: the aim of (4) is to influence broader cultural awareness and understanding of what science is, as well as what it can and cannot do. The motivation for this addition stems from the need to increase scientific literacy within religious circles while at the same time combating either the vilification or the idolization of science that happens within the same circles. Whether overly deferential or overly dismissive, mistaken views of science need to be addressed through more informed and accurate accounts of the actual and often messy ways in which scientists come to better understand the world.

Augustine famously asked whether any vestiges of God's triune character were to be found within creation itself (see, for example, Augustine 1998, Bk. 11.24ff.). Although many theologians since his time have seen fit to identify such vestiges, both in the external world of nature and in the internal world of the human psyche, this particular theological exercise suffered a withering critique in the early twentieth century at the hand of Karl Barth, who, in the face of dire political circumstances, resolutely rejected any attempt to argue for knowledge of God apart from revelation in Jesus Christ. Barth refused to accept the possibility of any *vestigia trinitatis* that could, alongside the biblical witness, be said to point to the triune nature of God (1975, 333–347). As Colin Gunton more recently noted, however, Barth's rejection of natural theology still leaves room for the idea that the world "speaks of the being of the one who made it" (1998, 144). Gunton rightly argued that this dynamic takes place within the context of prior belief, and not solely on the basis of scientific understanding. Accordingly, the narrative developed in these pages moves not from nature to God, but from God's relational character and activity in the world, as they have been variously expressed in the Christian tradition, to a critical and constructive re-expression of those ideas, in light of quantum entanglement. This approach takes its cue from Barth's more positive assessment of early *vestigium* arguments: "not that [the early theologians] tried to explain the Trinity by the world . . . on the contrary . . . they tried to explain the world by the Trinity in order to be able to speak about the Trinity in *this* world" (1975, 341; emphasis mine). I refer to this as a "theological *vestigium* argument" in order to distinguish it from any attempt to argue from instances of threefold relationality in nature to the trinitarian identity of God, which I refer to as a "natural *vestigium* argument."

Admittedly, a good deal of controversy still exists today over whether appreciative engagement with the sciences entails a rejection of the Bible and its authority for Christian life and thought. Against "either/or" views of the matter, Wolfhart Pannenberg has argued that theological reflection on the world *should* follow the biblical witness, not by turning it to stone but by employing "current knowledge of the world for a description of the divine work of creation, using the resources that are actually at hand" (1994, 117). Indeed, one can construct a theological view of the world in conversation with the sciences which, far from being anti-biblical, actually affirms and honors both the formation and the content of the biblical account of creation. The priestly writers articulated Israel's faith in terms of the most comprehensive account of reality in their time—Babylonian cosmology—and Christians today must seek to do no less (I owe this insight to Robert Russell, personal communication). Narrating the story of God's love for the world in Christ with the help of concepts and images from contemporary science is an endeavor that honors rather than denigrates the biblical accounts of creation.

Although the nonpersonal character of quantum entanglement makes it an unlikely metaphor for the God–world relation, considering it alongside Christianity's predominantly personalist collection of traditional concepts and images creates several opportunities for connection and reflection. First, nonpersonal metaphors situate human existence within the wider context of the inorganic world and universe, allowing us to conceive of our own existence as being with the grain of creation more generally. Second, they help to reestablish a sense of the connection between life and nonlife by promoting a vision of *all* reality as God's creation. Third, they force us to ask whether the physical world's existence, theologically considered, might be more than a necessary backdrop for our own. And, finally, they expose the parochialism of the presumption that the traditional starting point for Christian accounts of divine agency—human agency or intentionality—will be sufficient to guide our thinking about God's relation to creation more broadly. Even allowing for the unique importance of personal metaphors in the Christian tradition of naming God, it is a failure of imagination to suppose that God cannot relate to the physical world in a manner consistent with its own modes of existence. The unique power of entanglement as a theological metaphor is that it can also do justice to the internal relationality of human experience. It is not difficult to form an image in one's mind of a God who is a solid rock or a cool breeze to the afflicted. But the problem with rocks, streams, and mountains is that they live in the modern imagination more as external objects than as internal relations. The metaphor of entanglement has the potential to overcome this limitation insofar as it points to an internally, constitutively relational feature of the physical world, not to some otherwise interesting but fundamentally external and nonrelational aspect of it.

3   LIFE IN THE COSMOS

Over the past four centuries, the world accessible through the senses has become vastly larger and wondrously smaller at the same time, thanks to the invention of technologies such as the telescope and the microscope. We now find ourselves living at a meso-scale level of matter and motion, able to speak with some confidence about the content and structure of the *entire* universe—from quark to cosmos—but no closer to a common understanding of the significance of our own place within it (Primack and Abrams 2006). The unity of the world as the singular product of a divine act, so apparent to the Western mind before the advent of modern science, has retreated into the unfathomable enormity of the universe and into the equally unfathomable minuteness of the atom. Physicists in their newfound priestly role speak of science's ability to read "the mind of God" (e.g., Davies 1992), but such liturgies have limited power to rekindle in us a sense of the physical world as God's creation or a sense of our place within it. In the biological realm, too, the work of geneticists and neuroscientists threatens to undermine our sense of self-determination by providing an increasingly detailed account of how our genes and neurons influence who we are, what we do, and how we think (Peters 2002). Notwithstanding its many practical benefits, modern science has created something of a global cultural crisis by exposing the mythical (i.e., symbolic) quality of the traditional religious narratives that have heretofore grounded peoples' senses of identity and place in the world. If these older narratives provide little in the way of technological benefits, are they really telling us something useful or true about "the way things are"?

At the same time, the natural philosophers of our own era have been working to reinterpret and thereby reconstruct our world by renarrating the story of the universe and life within it in accord with the labors of modern science. The sharp metaphysical dualism of the preeminent early modern thinker René Descartes (1993), in which the *res cogitans* (the mental-spiritual world of human volition and freedom) must be distinguished from the *res extensa* (the mechanistic world of inert, lifeless bodies, including all plant and animal life), has given way to the cosmic, developmental organism of thinkers such as Alfred North Whitehead (1957) and Pierre Teilhard de Chardin (1975). Of great interest to theologians, these new stories emphasize and celebrate the special features of life—novelty and chance, dynamism and interconnection, relationality and wholeness—not just in humanity but throughout the cosmos. Instead of limiting life to humans, as Descartes did, thinkers like these paint pictures of the universe in which life and nonlife, spirit and matter, blend seamlessly into one another.

Theologians, for their part, have begun to take the cosmic-evolutionary history of life seriously in their own theological accounts of the world as God's creation (see, for example, Barbour 2002; Edwards 1999; Giberson 2008; Haught 2000; Hefner 1993; Korsmeyer 1998; Miller 1999; Russell, Stoeger, and Ayala 1998; Peacocke 1993; Peters and Hewlett 2003; Polkinghorne 2001a; Schmitz-Moormann 1997; Zycinski 1999). Yet the strength of these new theological stories is also their weakness. By focusing on the precarious edifice of life—that wondrously fragile and possibly rare phenomenon within the universe we find ourselves to be—they fail to register the significance of the vast bulk of the universe that takes the form of nonliving matter. Does the "dead" part of the universe as we have come to know it through the sciences manifest any sort of relationality, dynamism, or novelty? Does it exhibit a kind of "life" of its own? Or is it just the same old inert, Newtonian stuff blindly bumping into itself over and over again?

The two pillars of twentieth-century physics—special relativity and quantum theory—point in somewhat different directions on this question. Einstein's special theory of relativity accounts for physical motion by treating time and space together as a single, four-dimensional manifold (commonly referred to as "space-time"). This profoundly relational approach to space and time has been repeatedly tested and resoundingly confirmed (see Will 2006, section 2.1.2; for a survey of the ways in which special relativity and other areas of physics elucidate different relational aspects of the physical world, see Polkinghorne 2010a; Nicolaidis 2010). Einstein's theory of gravity—his so-called general theory of relativity—builds on special relativity and has also been confirmed to a high degree of precision (Will 2006, passim). However, apart from some additional and not insignificant complexities, Einsteinian relativity preserves two key elements of the classical perspective. In a relativistic world, (1) all causal influences must propagate from one neighbor to another at some finite speed (i.e., the so-called "locality" of causal interaction), and (2) all points in space-time are fully individuated inasmuch as each has its own definite characteristics wholly unto itself.

The other pillar of modern physics, quantum theory, is used by physicists to characterize and understand the basic constituents of matter at the atomic and subatomic levels. It is blatantly "nonclassical." Classical physics began with the assumption that all dynamical properties are capable of continuous variation over a range of values. Consider the property "speed." No matter how fine-grained the analysis, the speed of any physical object will always be seen to vary smoothly from one value to another. Quantum physics rejects this classical assumption and assumes instead that all dynamical properties are "quantized" at some basic level, i.e., capable of taking on only certain discrete values called *quanta*, singular *quantum* (as though an object could travel at 10 or 15 kilometers per hour but nothing in between). And whereas relativity theory sticks with the locality and reductionism of classical physics, quantum theory points toward the nonlocality and wholeness of physical processes in its treatment of physical systems composed of multiple objects. The relativistic and quantum perspectives are not obviously reconcilable with one another, although proponents of "string theory" controversially claim to have found a unifying framework (cf. Chiao 2004, 254). Some theologians have attempted to incorporate these two perspectives into their organic visions of the cosmos, but characterizations of the physical world as the product of God's *life-giving* presence are inevitably constrained by where they begin. Can the physical world ever surface as anything other than background in a story that takes the special features of "life" as its starting point? If through its newly articulated cosmic-evolutionary character life once again seems worthy of the title "creation," what of the universe in its totality? Throughout these pages I aim to demonstrate that physics provides its own unique storehouse of concepts and images for reenvisioning the universe theologically—*qua* physical object—as creation (for important efforts in this direction, see Heller 1996; Hodgson 2005; Murphy, Althouse, and Willis 1996; Polkinghorne 2007, 2010b; Russell, Stoeger, and Coyne 1988;

Russell, Murphy, and Isham 1996; Russell et al. 2001; Russell 2006; Sobosan 1999).

4 WHY QUANTUM PHYSICS

Two particular aspects of quantum theory have received considerable attention among theologians in recent decades. The first, the idea of "complementarity," refers most generally to the impact of the experimental apparatus or context on which of an object's properties can be meaningfully considered and observed. For example, by configuring some particular measurement apparatus in one way, one can observe a quantum object's characteristically "quantum" behaviors, but one cannot say how the object navigates its way through the apparatus to produce such behaviors. By configuring the apparatus differently, one can see precisely how the object moves through the apparatus, but one can no longer observe any of its distinctively quantum behaviors. These different types of observation and information are "complementary" in the sense that one must choose what one wants to measure. It is simply not possible, according to quantum theory, to observe all of the quantum object's properties simultaneously. Theologians have appealed to complementarity to warrant their use of dialectical arguments as well as the possibility of peaceful coexistence between science and religion (e.g., see Barbour 1997; Kaiser 1996; Mackinnon 1996; Weisskopf 1992). The notion of complementarity as it has been used by quantum physicists is surely a profound insight into the complex and inevitably partial ways in which humanly derived categories overlay reality, but its use—whether in science or theology—always puts one at risk of making facile paradoxical assertions.

The other widely presumed feature of quantum theory can be summed up in the phrase "causal indeterminism," by which is meant the apparently random character of physical events at the quantum level (such as the decay of a radioactive atom). Most physicists and philosophers take the apparent randomness of quantum events to reflect the ontologically underdetermined character of physical processes in general. There simply is no natural cause, according to the standard interpretation, that can be invoked to account for why, say, one radioactive atom decays now rather than several days later. If correct, this view implies a world of genuine chance and novelty rather than the Newtonian world of lockstep determinism. Robert Russell and others have appealed specifically to the idea of quantum indeterminism in their attempts to mount a theological argument for the possibility of noninterventionist, objectively special divine action in the world (see, for example, the essays by Murphy and Tracy in Russell, Murphy, and Peacocke 1995; the essays by Clayton, Tracy, Ellis, and Russell in Russell et al. 2001; and Russell, Murphy, and Stoeger 2008). The force of the argument hinges crucially on whether or not the divine determination of an otherwise underdetermined physical process ought to be understood as an "intervention" on God's part into the causal structures of the world (see, for example, the debates in Russell et al. 2001; Russell, Murphy, and Stoeger 2008).

The appeal of quantum indeterminism with regard to the articulation of a theologically and scientifically plausible (i.e., noninterventionist) mechanism for objectively special divine action stems from the presumption that divine and creaturely power must compete with one another in a "zero-sum game" at the level of ordinary events (Wegter-McNelly 2008). Whenever and wherever God acts to steer the world counterfactually toward a particular future, it cannot also be acting through its own causal structures to steer itself (and vice versa); otherwise one risks the conceptual confusion of overdetermination. Beneath this basic assumption lies an even more fundamental theological intuition, namely, that God really is present to the world in such a way as to steer it toward a particular future. When combined with a sensitivity toward the integrity of scientific explanation, this linchpin of the classical Christian understanding of God's relation to the world requires God's activity in the world to make a counterfactual difference while remaining hidden from scientific investigation. Divine presence thus takes the form of objective, special (i.e., counterfactual) action without thereby becoming interventionist.

Another prominent approach to divine action in the contemporary theological literature—that of *kenosis*—similarly embraces the hiddenness of divine action from scientific investigation (e.g., see Polkinghorne 2001b). Advocates of this approach, however, question the theological viability of a God whose activity is tied to the underdetermined interstices of quantum processes. The self-emptying power of divine love, they contend, offers a surer defense against the problem of evil and a more solid theological basis for eschatological hope. In Chapter 6, I take my cue from the kenotic theologians in characterizing God's presence in natural processes as the freedom-granting power of self-relating and other-relating love. I argue, however, for a plerotic rather than kenotic account of divine agency on grounds that a plerotic view of divine presence as "entanglement" affords greater overall conceptual coherence to the theological perspective. In contrast to both the quantum and kenotic theologians, I also question the assumption that God counterfactually steers the world toward a particular future by rethinking a variety of theological issues from the vantage point of God's entangled relationship with creation.

In comparison to complementarity and indeterminism, quantum entanglement has been the focus of relatively little theological discussion to date. But just as there has been a growing recognition of the theoretical and practical significance of entanglement within the wider scientific and philosophical communities, theologians have begun to wrestle with its metaphorical and metaphysical significance (see Russell 1988, 2001; Sharpe and Walgate 1999; Simmons 1999, 2000, 2006; Buxton 2005; Conner 2006; Voskuil

2007; Wegter-McNelly 2007; Polkinghorne 2010b). Unfortunately, even the more recent contributions to this discussion have done little to engage the growing body of philosophical literature on entanglement. It is worth recalling in this regard that within a decade of the appearance of quantum theory, Erwin Schrödinger, one of the theory's chief architects, had already come to the essentially philosophical conclusion (1935) that entanglement, more than indeterminism or complementarity, constitutes *the fundamental feature* of this physical theory that distinguishes it from all previous ones. As we shall see, much recent philosophical reflection affirms Schrödinger's initial assessment.

It may nonetheless be tempting to dismiss entanglement as just another quantum oddity that is, for all practical purposes, irrelevant to the ways in which we experience the world. Consider some of the bolder claims quantum physicists commonly make about the nature of physical reality: most if not all of reality exists in fuzzy, indefinite states called "superpositions" (which affect the course of events but which we never directly observe); quantum particles spread out like waves when we are not looking at them, but they always show up in experiments as particles with well-defined locations; and finally, the human act of observing a quantum process inevitably alters how that process proceeds, relative to how things would have gone if no observation had been made. At our own level of existence, such "insights" seem largely irrelevant. We always find physical objects in one definite state or another. We never see particle-like objects acting like waves or vice versa. And the fundamental limit on measurement imposed by Werner Heisenberg's famous uncertainty principle still allows plenty of room for precise prediction and measurement.[4] Why not add entanglement to the list of strange but safely ignorable features of the microphysical world?

Those unfamiliar with quantum theory can be forgiven for continuing to view the physical world as a giant collection of blindly bumping billiard balls. They may also be forgiven for continuing to think of physics as uninteresting for, and uninterested in, the subject of relationality. The difficulty here is that the success of quantum theory has led many physicists and philosophers to argue that the world we live in *cannot* be the same old Newtonian world of fully distinguishable and separable objects. In fact, although it is not yet widely appreciated outside scientific circles, quantum theory has already made a significant impact on contemporary culture inasmuch as quantum technologies—for example, the transistor and the laser—are crucial components of the ongoing revolution that is the Age of Information. As our awareness of the relevance of quantum theory to a host of scientific, technological, and even philosophical problems increases—from ultrafast computing to nanotechnology (Milburn 1996), perhaps even to the nature of consciousness itself (Hameroff and Penrose 1996)—the impact of this remarkable theory on our understanding of ourselves and the world around us can only deepen.

Entanglement as a theological metaphor must be malleable enough to cover the particularities and peculiarities of various instances of theological appropriation. In some cases, personal considerations will alter how the concept must be understood. The asymmetry between God as creator and the world as creation means that the phrase "God is entangled with the world" cannot mean exactly the same thing as "the world is entangled with God" (or, of course, "two photons are entangled with one another"); the metaphor will need to be nuanced and adapted in light of the particular theological context. That said, throughout the book I hew as closely as possible and for as long as possible to the meaning of entanglement in its physical context. Reshaping the term too quickly in the service of theology runs the risk of weakening the coherence and integrity of the overall vision, which hinges on the claim that entanglement in the physical world can be understood, from a theological perspective, as a reflection of the entangled nature of God (i.e., as a "theological" *vestigium*). On the other hand, I attempt to face squarely the "is not" of entanglement when disanalogous aspects come to the fore. The power of entanglement as a theological metaphor lies centrally in its potential to broaden and deepen Christian theological speech regarding the complex relationships among divine immanence, divine transcendence, creaturely immanence, and creaturely transcendence. As a product of human thought and speech, this metaphor brings together a theological description of an intrinsically relational God with a scientific description of an intrinsically relational world while preserving the distinctiveness of each perspective.

5   WHY MATHEMATICS

Most books written for theological audiences contain little or no math, and whatever does appear is commonly relegated to appendices. In the present case such an arrangement would communicate two misleading messages: first, that the math associated with entanglement is hopelessly hard, and, second, that the math is irrelevant to the larger theological agenda. It is not. Again, I urge the nonscientist reader to refuse to accept the claim that the details of Bell's theorem lie beyond his or her ken. Anyone with a basic grasp of high school algebra, a faint memory of trigonometry, and a modicum of patience can follow the derivation—which is surprisingly short, given its importance in the history of modern physics. Bell's original version was just a few pages long. Mine is longer, but only because I take the time to explain each step in detail.

Three additional reasons weigh against relegating the math to appendices. First, a book that combines the conceptual worlds of physics and theology needs to include enough detail for experts in each field to judge for themselves whether or not the connections are well motivated. This cannot happen without a review of the math. Additionally, working through the math makes obvious, in a way that prose alone cannot, the fact that none of the interpretations of quantum entanglement considered in Chapter 5 is somehow located *within* the mathematical formalism[5] of quantum theory. Saying what a scientific theory means for one's understanding of the world is no less an interpretive endeavor than saying what a theological doctrine means for one's understanding of God (notwithstanding the fact that theological reflection may implicate more levels of interpretation; see Clayton 1989, 101). Judging the merits of a particular interpretation of quantum entanglement is a task that must be deeply informed by, but never follows simply from, one's understanding of the relevant science. Here again the viewpoint of an interdisciplinary metaphorical theology comes into view. For a theological metaphor like the one developed in this book to have merit, it must be closely related to a plausible interpretation of the physics, even though interpretive fidelity to the science cannot be the only consideration (others will include resonance with human experience, relation to biblical and previously existing theological categories, the need for coherence across the theological themes, etc.).

Second, as Robert Russell has noted (personal communication) any contribution to the ongoing theology-and-science discussion that wants to be taken seriously cannot rely solely on secondary "translations" of the science. Interdisciplinary work demands an interpretive awareness that can only come through engagement with the primary literature. In the case of quantum entanglement, familiarity with the relevant math provides an important check against finding one's own view confirmed somewhere in the secondary literature and then claiming the mantle of scientific authority. The inclusion of mathematical material in these pages is meant to give both power and pause to those who might wish to venture their own opinion about what quantum theory "really" says about the world. Knowing the math will not settle the interpretive disputes, but it can keep these disputes in perspective by foregrounding the constructive nature of all interpretation and, when necessary, exposing interpretive malpractice.[6]

Third, mathematically accessible and interpretively nuanced derivations of Bell's theorem are unfortunately difficult to come by. Watching this theorem unfold over the course of a few pages can, I hope, bring the reader a sense of joy and accomplishment, as well as a desire to join with those in the larger theological academy who are working to find new ways of engaging and learning from the sciences. A basic conviction that shapes my own theological perspective, as the reader will no doubt have already discerned, is that theologians must learn to listen and speak across the cultural chasm that separates theology from the sciences. The presence of a slow and careful, physically intuitive derivation of Bell's theorem within the larger narrative of this book honors that commitment and provides at least one model for theologians who wish to engage the sciences with integrity.

6 ENTANGLEMENT—A THEOLOGICAL METAPHOR?

In English the verb "to entangle" has a variety of related meanings, all of which are strongly negative. To be entangled is to be trapped, caught up in something from which escape is difficult, or bound in an embarrassing or compromising relationship. Entanglements ensnare, fetter, immobilize. Consider a representative sampling of headlines employing the word: "Drowned Man Entangled in Wheelchair," "Schools Entangled in Red Tape," "Entangled Economies Cause Friction," "British Socialite Entangled in Scandal," "Nepal Entangled in Temple Row," "Skydivers Die as Parachutes Become Entangled." When it appears in various English translations of the Bible, the word likewise has negative connotations. The NRSV, for example, opts for "entangled" as a translation of ἐμπλέκεται in 2 Timothy 2:4 and ἐμπλακέντες in 2 Peter 2:20, (both from ἐμπλέκομαι: to be mixed up in or involved in). In 2 Timothy a good Christian is compared to a soldier who avoids becoming "entangled in everyday affairs" so as to have the freedom to focus on pleasing the enlisting officer. In 2 Peter those who turn away from the teachings of false prophets but again become "entangled in them" are like dogs that turn back to their own vomit and pigs that wallow in the mud after being washed. What then is the rationale for speaking of God and the God–world relation as "entangled"?

The first point to be made is that the words "entanglement" and "entangled" carry a rather different meaning in the context of quantum physics. Erwin Schrödinger, one of the inventors of quantum theory, appears to be the source of these words in that context. In a response to EPR's 1935 paper, he pointed out that once two physical systems have interacted, they can no longer be assigned separated mathematical descriptions in the quantum formalism. The formalism instead requires that the two systems be represented by a single mathematical construct. This, he said, is not "*one* but rather *the* characteristic trait of quantum mechanics, the one that enforces its entire departure from classical lines of thought. By the interaction the [objects' separate mathematical representations] have become entangled" (1935, 555; italics in original; cf. Schrödinger 1983). The closest word in Schrödinger's native German is *Verschränkung*, which can mean folding, crossing, clasping, interleaving, or interconnection. The connotations in German are not as negative, but "entangled" is the English word Schrödinger chose to use in his 1935 essay (he wrote scientific papers both in English and in German) and it has stuck. In 1991 the *Oxford English Dictionary* added the following definition to its entry on entanglement: "*Physics.* A correlation between the states of two separate quantum systems such that the behaviour of the two together is different from the juxtaposition of the behaviours of each considered alone." Although its more traditional meaning in English has by no means been supplanted, the term's semantic range is increasing in ways that complicate the older meaning.

In light of such semantic flux, I suggest that the term "entanglement" is ripe for theological consideration. The investigation of this idea in the quantum context developed in Chapters 5 and 6 reveals its potential for shouldering a complex and multivalent set of theological connotations. My own interest in the theological potential of entanglement is shaped in particular by recent literature on divine suffering (e.g., Moltmann 1993a; Fiddes 1992; Placher 1994; Polkinghorne 2001b; Chung 2002). If the cross emblematizes God's embrace of creaturely suffering, then perhaps it is possible to link God's entanglement *in suffering* to God's desire to become entangled *with the world* by creating it. Divine entanglement in and with the world will have both positive and negative aspects. The former can be articulated in terms of the creative work of a relational God, for whom "entanglement" becomes a sign of the inner-divine mutuality that grounds the possibility of healthy difference and right relationship among creatures as God's creation. The negative aspects can be associated with, if never satisfactorily explained by, creaturely resistance to right relationship. Such resistance leads to the breaking of right relationships or to the unhealthy acceptance of harmful ones. The divine embrace of suffering that happens in response to harmful creaturely entanglements can itself be thought of as a loving kind of entanglement that remains with the harmed, even to the point of death. God's entanglement with and in creation is not an overpowering charge-taking but rather an act of divine empowerment that enables creaturely freedom. In choosing to relate to creation outside Godself (*ad extra*), God establishes the possibility for creation to be and become itself, freely, in relation to its creator. This, I will argue, is what it means for God to be entangled.

2 Relationality in Contemporary Theology

The autonomous, self-constituting person, the much vaunted "I" of the Enlightenment, seems now to have had its day. The solitary cogitating mind, formulating itself and truth from the ground up—the icon of modernity—turns out to have been an illusion born of privilege. In actuality, it was typically only those with sufficient means to pay others to take care of "life" who could deceive themselves into thinking that they were the sole determiners of their own future. Self-determining modern "man" was in fact nothing apart from the relationships he relied upon but failed to see, the various quills with which he wrote his own story: the wife, the housekeeper, the gardener, the environment, the servant, the slave, the worker, the child. In our time, however, in what is a late modern or perhaps already postmodern moment in the West, the image of the "I" as prior to and independent of the "We" is receding in favor of a new vision that sees the "I" as a refracted "We." And, as others from non-Western cultures continue to remind those of us in the West, community as the source of individuality is not actually a new idea. In truth, it is not even a new idea in the West.

Consider the early churches of the first four centuries. Out of the struggle to articulate their understanding of Jesus's identity there emerged a relational view of God which eventually came to be codified in the doctrine of the Trinity.[1] One of the ironies of Western individualism, especially for Western Christians, is that despite the dominant and dominating cultural place of Christianity in the West since the time of Constantine, the church's early relational view of God has until recently had relatively little impact on subsequent Western philosophical or theological discussions of the triune nature of God or of the human personhood as *imago Dei*. Today the idea that the individual emerges out of community can sound as odd to Western Christians as it might to any other Westerner. This is due in large part to the fact that the classical theologians of the Latin West emphasized the primacy of God's oneness—*una substantia*—as a guard against tritheism, whereas those in the Greek-speaking East gave greater weight to God's plurality—*tres hypostaseis* (albeit in subordinationist terms that positioned the Father above the Son and Spirit).

These ancient layers of history, going back to the Enlightenment and before, lie at considerable remove from the twenty-first century, yet they are important touchstones for our consideration of recent theological interest in relationality. We begin with an analysis of recent cultural dynamics that have led to an upsurge of theological interest in the idea. Following this, we consider the importance of three contemporary theological movements that have spurred interest in relationality. The theological nuances of the term "relationality" occupy the next section, which in turn leads to a broad-ranging and selective survey of the theological literature on relationality. Finally, we analyze in greater detail the relational accounts of creation given by two contemporary theologians: Wolfhart Pannenberg and Sallie McFague. Readers wanting to explore the theological literature on relationality in greater depth may consult the Appendix for a more comprehensive list of relevant late twentieth-century theological monographs.

1   THE CULTURAL TURN TO RELATIONALITY

A variety of cultural factors lie behind the recent theological turn to relationality, perhaps the most important of which is the advent of the environmentalist movement. In the mid-twentieth century writers such as Aldo Leopold, author of *A Sand County Almanac* (1949/1987), and Rachel Carson, author of *Silent Spring* (1962/1994) sounded a clarion call to value all life on Earth and act in the face of its perilous condition. Ecology rose to prominence as a scientific field during the same time, introducing the wider culture to terms such as "ecosystem," "food web," and "biodiversity." Today environmentalists continue to highlight the impact of human practices on fragile ecosystems, as well as the potentially global effects of local practices. The environmental movement, which now stretches around the globe, has added its own wisdom about the interrelatedness of life to those insights already gained from evolutionary biology. As our understanding of humanity's place in the vast web of life grows, people continue to look for ways of living more lightly in the world. In the theological realm, these events have been paralleled by the rise of "eco-theology," now a significant voice in the larger world of theology (e.g., Bouma-Prediger 2001; Cobb 1992; Deane-Drummond 2008; DeWitt 2007; Edwards 2006; Hallman 1994; Hart 2006; Rasmussen 1996; Ruether 1992). Cultural interest in ecological issues has indeed been an important factor in generating theological interest in relationality, but it has also focused the theological discussion on a particular type of relationality—that which characterizes what we call *life*. As I will argue in the following, focusing exclusively on this type of relationality is problematic for a Christian who wants to be able to see the entire world—including the inanimate world—as creation.

A second cultural factor behind contemporary theological interest in relationality is the continuing power in Western cultures of the mythos of individualism, a perspective which initially took hold in the European Enlightenment and continues today in various forms and locales. In the U.S., the "self-made individual" is still the default posture of politicians on the campaign trail. Any public figure who wants to suggest that Americans should put the "common good" ahead of personal gain is likely to be tarred a socialist or a communist, or even a fascist. Recall the 2010 exchange between the organization Jewish Funds for Justice (JFJ) and political commentator Glenn Beck, in which the latter asserted that JFJ's promoting the "common good" was both dangerous and naïve. Putting the common good first was, he said, what led to the Nazi death camps. Entrepreneurialism, private property, limited government—each of these values is trumpeted by the religious right in the U.S. as compatible with Christian ethical principles. Progressive religious communities respond by pointing to the communitarian ethic that permeates both the Hebrew Bible and the New Testament, only to be dismissed as left-leaning radicals who put politics ahead of faith. At least one current political commentator, however, recognizes the powerful critique of individualism contained within the biblical ethic. After the 2010 earthquake in Haiti, Jon Stewart of *The Daily Show* responded to Pat Robertson's televised comments about the Haitians deserving their fate by shaking a large Bible at the camera and asking Robertson, "Have you read this book?" Stewart went on to quote several biblical passages extolling God's compassion and care for the brokenhearted. In the theological sphere, influential figures such as Stanley Hauerwas (2000) and John Milbank (2005) have argued against any easy alliance between Christian principles and the Enlightenment's story of self-sufficiency, urging Christians to reclaim the ethical visions of the Bible and early Christianity. Reacting to these ongoing cultural debates over the value of individualism, the contemporary theological discussion about relationality often includes a strong political component.

Finally, we turn to the near dissolution of Christendom in the West, first in Europe and now in the U.S., and how this development is reshaping the religious landscape. As the demographics of the European–North American axis continue to diversify, the plurality of religions has become an ever-more concrete reality for its inhabitants. Modernity, it turns out, has caused not the death of religion but rather the proliferation of religious options. National discourses shift accordingly, as when, for example, the neutral tone of the "disestablishment" clause of the First Amendment to the U.S. Constitution is reinterpreted through the lens of "religious pluralism." With the hegemonic power of Christianity on the wane in Europe and North America, Christians in these regions are increasingly put on the defensive with regard to their right to speak *qua* Christians in the public sphere. How does Christianity understand itself vis-à-vis the world's other religions when it no longer occupies a position of cultural dominance? How should Christians seek to participate in the common political and social life of increasingly multireligious cultures? Questions like these give the theological discussion of relationality an existential dimension: How can people of different faiths live together peacefully and cooperatively, yet without denying their differences? Let us now examine several important theological responses to broader cultural developments that have contributed to increased interest among theologians in the category of relationality.

2 THE THEOLOGICAL TURN TO RELATIONALITY

First let us consider the decreasing plausibility of the traditional Western "substance" view of reality. Both the view itself and its decline have had a profound impact on Western theology. The West inherited the idea of "substance" as a basic metaphysical category from Aristotle, who took reality to be populated by enduring, individual objects whose relations to each other are unrelated to their own essences or identities (relations for Aristotle were famously "accidents" rather than "substances"; see Brower 2009; Shults 2003, chap. 1). The emergence of relationalist alternatives in the early and mid-twentieth century—most notably the process-relational perspectives of the philosopher Alfred North Whitehead (1957) and the theologian Charles Hartshorne (1948)—has provided theologians with a new set of tools for tackling a host of long-standing theological problems (e.g., the problem of evil, the question of immortality) and interpreting social issues (e.g., climate change, economics, sexuality). The process theological perspective is fundamentally relationalist and anti-Aristotelian in the sense that it regards all "things," including God, to be themselves primarily by virtue of their relations to other "things" (this is the so-called process doctrine of "internal relations"). For more than four decades now, process theologians have been exploring the process-relational view of being and becoming for the sake of theology's intelligibility (Wegter-McNelly 2000) in an age of often overwhelming social, intellectual, and religious flux (for important early process writings addressing the topic of relationality, see Birch 1972; Clarke 1973; Fontinell 1979; Fowler 1979; Hurley 1979; Suchocki 1989; Bracken 1984; Cobb Jr. 1984, 1986; Loomer 1984; Reeves 1986; Keller 1987b; Sullivan 1987; Howell 1989). The decline of substance metaphysics—welcomed by some but contested by others—is a cultural and philosophical transition of paramount significance in and for the West. As a result, numerous theologians across the theological spectrum have begun to reexamine and recast the wisdom of their own traditions in light of process-relational categories.

A second important theological factor driving the theological turn to relationality has been the emergence and growth of feminist theology alongside second-wave feminism. Feminist theologians have registered a sharp critique of Western, androcentric theology for associating males predominantly with (theological) reason and females predominantly with (religious) emotion, and above all for encouraging and enforcing the subordination and silencing of women at home, in the church, and in society throughout history. At the heart of early and more recent feminist theological scholarship are the categories of relationship and relationality. Feminist theologians have generally sought theologies and anthropologies that are less oppositional and more cooperative, less individualistic and more communal, less glorying in the achievement of the one and more valuing of the contributions of the many. These thinkers have been centrally responsible for introducing the ethical and ontological categories of "right" and "wrong" relation into the larger theological discussion. Their power continues to manifest itself within feminist theological writing across the personal, familial, professional, societal, national, ecological, and spiritual realms (for a representative selection of feminist theological writings on the topics of relationship and relationality, see Russell 1979, 1981; Ernst and Ernst 1982; Raitt 1982; Walsh 1982; Keller 1985, 1987b, 1989, 1997; Suchocki 1985; Caton 1987; Cooey, Farmer, and Ross 1987; Krobath 1988; Yeager 1988; Hampson 1989, 1993; Howell 1989; Hunt 1989; Welch 1990; Grey 1991, 1999; Johnson 1992; Grigg 1994; Herrington 1996; Jakobsen 1997; Neuger 1999; Spalding 1999a, 1999b; Lowe 2000; Coakley 2009; for a provocative and erudite work on relationality that combines science studies, feminist theory, and quantum physics, see Barad 2007, esp. chaps. 3 and 4; I thank Catherine Keller for drawing my attention to Barad's work— unfortunately, publication deadlines prevented me from engaging her work at length in these pages).

A third important theological factor behind the rise of relationality in contemporary theology has been the reinvigoration of the doctrine of the Trinity that began in the early and middle parts of the twentieth century with the work of two highly influential theologians: Karl Barth and Karl Rahner. Their initial work on the relationality of God was, after a period of quiescence during the middle of the century, actively embraced by a group of theologians who are sometimes referred to as the "neo-trinitarians." In addition to mining the depths of Barth and Rahner, these theologians have gone beyond Barth and Rahner to reclaim and creatively reappropriate the Cappadocians' early insights on divine plurality. Here I briefly outline these developments, beginning with Barth and Rahner's return to the Trinity, then moving to the more recent expansion of this agenda by the neo-trinitarians.

To many early and high modern thinkers in the West, the Trinity looked very much like an outmoded, mystical concept ill-suited to the *aufgeklärten* theology of a rationally sophisticated age. In 1936 Swiss Reformed theologian Karl Barth took issue with this assessment in his characteristically forceful way, beginning his magnum opus, the *Church Dogmatics*, with a detailed explication of the doctrine of the Trinity. In doing so he signaled his intention to reclaim the particularity of the Christian God as an antidote to what he saw as Christianity's abandonment of its distinctive theological voice during the eighteenth and nineteenth centuries. Barth grounded his explication of the doctrine of the Trinity in the *self*-revelation of God: God is revealer, revelation itself, and the revealing of that revelation (1975, 314). Using an intensely Christological and biblically oriented approach, he insisted that the Christian account of God must be based specifically and solely on the revelation of God in Jesus Christ or, in more abstract language, on "the freedom of God to differentiate Himself from Himself, to become unlike Himself and yet to remain the same" (320). This dialectical vision of divine relationality found its source within the (to Barth, incomprehensible) tri-unity of God, whose nature he described as "three distinctive modes of being of the one God subsisting in their relationships one with another" (366). Barth did not go so far as to equate divinity with relationality—this, as with any effort to pin human concepts on God, would have seemed to him the height of hubris—but his rich explication of the distinctiveness of trinitarian thinking remains an important touchstone for the contemporary discussion.

The German Roman Catholic theologian Karl Rahner was occupied with a very different project, namely, that of bringing his own tradition into a more constructive relationship with modern thought, especially in relation to the Kantian and existentialist traditions. Like Barth, and much indebted to him, Rahner foregrounded the doctrine of the Trinity in his own theological writing. Also like Barth, he saw divine revelation as God's act of relating through giving *Godself* to the creature, rather than God's act of imparting to the creature mere *knowledge about* God. Rahner was concerned that the doctrine of the Trinity had become overly speculative and removed from the existential realities of our own lives. He argued that Christian theology must not conceptualize the immanent Trinity (God in Godself) apart from the economic Trinity (God in history). Although Christian theology needs both ways of talking about God, one should avoid speaking of the immanent and economic Trinities as though they were two different Gods. The economic Trinity, said Rahner (1970, 22), is the immanent Trinity, and vice versa—a maxim that has come to be known as "Rahner's Rule." By implication, the relationship between Jesus and God attested to in the biblical witness must be understood to reflect who God is in Godself. God does not shed unrelatedness or isolation in the act of creation or incarnation, and neither act is to be understood apart from the other (cf. Rahner 1976, chap. 6, esp. 197). Like Barth, Rahner conceived of relationality within God as a key feature of the Christian account of divine oneness. Relationality as such did not command the attention of either thinker, but their accounts of divine relationality paved the way for the more radically relational agenda that has emerged more recently.

In the wake of Barth and Rahner's recovery of the doctrine of the Trinity (see also Jüngel 2001, the first German edition of which appeared in 1965), a number of late twentieth-century theologians began to explore this doctrine anew. The list of so-called "neo-trinitarians" is long, but some of the more prominent contemporary voices include (roughly in chronological order): Jürgen Moltmann, Wolfhart Pannenberg, Leonardo Boff, Catherine Mowry LaCugna, Colin Gunton, Stanley Grenz, John Zizioulas, Walter Kasper, Joseph Bracken, Elizabeth Johnson, Robert Jenson, Ted Peters, S. Mark Heim, Kathryn Tanner, and David Cunningham. Broadly speaking, these thinkers take the central and lasting contribution of the doctrine of the Trinity within the history of Christian thought to be the fundamentally relational view of God that it established and preserved. Pannenberg's thoughts on the matter are representative:

> Any derivation of the plurality of the trinitarian persons from the essence of the one God, whether it be viewed as spirit or love, leads into the problems of either modalism on the one hand or subordinationism on the other. . . . the relations between the persons are constitutive not merely for their distinctions but also for their deity. . . . The task [is] to envision as such the unity of the divine life and work that is manifest in the mutual relations of the Father, Son, and Spirit. This requires a concept of essence that is not external to the category of relations. (1991, 298, 323, 335)

The neo-trinitarians' affirmation of the idea that relationality is ontologically basic to the divine being constitutes a significant departure from the patterns of thought that have dominated Western theology, which from Augustine to Barth and Rahner prioritized the oneness and aseity of the divine subject. Behind this departure lies, on the one hand, the influence of Vladimir Lossky's assertion (1957, esp. chaps. 1–3) that a clear difference could be found between "Eastern" and "Western" forms of trinitarianism—with the former focused sharply on relationality—and, on the other hand, John Zizioulas's more recent reinterpretation (1985, 2006) of the theological heritage of the Greek-speaking East around the notion of "communion" (for a concise survey of the revival of trinitarian theology in the twentieth century, see Coakley 2010).

Especially influential in neo-trinitarian circles has been Catherine Mowry LaCugna's magnum opus, *God for Us*. This theological and historical *tour de force* revisits the theology of the Cappadocians—Basil of Caesarea, Gregory of Nyssa, and Gregory of Nazianzus—in order to reclaim and rehabilitate their account of divine relations for the West. At the heart of the Eastern view LaCugna saw the notion of *perichoresis*, a "being-in-one-another, permeation without confusion" (1991, 271). This idea was first articulated by Gregory of Nazianzus with regard to the mutual interdependence of Christ's two natures and then applied several centuries later by John of Damascus to the dynamism and interpenetration of the trinitarian persons. LaCugna developed her interpretation of the Greek perspective in order to bolster an argument for the world-affirming character of the doctrine of the Trinity. The relationality of the God manifested in Jesus Christ, she said, is not an abstract idea but the lodestar of Christian life.

God in Godself is fundamentally *for us*. Zizioulas expresses the same view in ontological terms: "If communion is conceived as something *additional* to being, then we no longer have the [correct] picture. The crucial point lies in the fact that being is *constituted* as communion" (1985, 101; italics in original). For the neo-trinitarians, the doctrine of the Trinity cannot be construed primarily as a doctrine about who God is apart from creation; the communion of divine being includes, through the loving divine will, that which is not God, i.e., the creation. This relational view of God has deep roots in the Christian tradition, as we have seen, and yet in recent decades has flowered in a new way.

3 RELATIONALITY—WHAT ARE WE REALLY TALKING ABOUT?

I have already alluded to some of the tones and nuances given to the term "relationality" in recent theological writing, but we must pause for a moment to consider more carefully its range of theological meaning because one of the aims of this book is to broaden that range. What do contemporary theologians typically mean when they invoke the term "relationality"? As theological speech about God, the term "relationality" is frequently used to evoke dynamism and mutuality, a unity that incorporates difference, an open and welcoming orientation toward "the other." As theological speech about humanity, it commonly signals the complex interactions people have with each other and the world around them, as well as their connections to the past and future. Understood as a virtue, "relationality" is the truth, goodness, and beauty of being as "being with." It is the mutuality of "communion."

From a theological perspective, therefore, the most interesting relations are not those that follow simply from the logic of multiplicity (e.g., if A and B are distinct concepts, then there is necessarily some relation between them). Rather, theologians are interested in those relations that carry the meaning of our lives (for thoughtful accounts of the concept of "relation," see Wildman 2010; Welker 2010; Coakley 2010; for a striking, manifesto-like collection of essays that aims to steer the contemporary theological discussion in the direction of multiplicity and relationality, see Keller and Schneider 2010). I have a certain relationship with the handles on the cupboards in my kitchen—they sometimes twist my fingers uncomfortably when I pull them. Being the kind of person who appreciates things that are thoughtfully designed, this is a source of minor irritation in my life. But my relationship with these handles means relatively little to me, much less, for example, than the one I have with the walnut hutch that stands at one end of the dining room. It was made by pioneer settlers in the Midwest and refinished a century later by my parents, both of whom are now gone. And for all the warmth I feel toward this rather crude and astonishingly heavy piece of furniture, my relationship with it means little in comparison to those I have with my spouse and my son. The relationships we have with family, friends, and coworkers are the daily context and often the subject of our deepest joys and sorrows.

The best of our relationships empower us to make meaning out of the diverse circumstances of our lives. They occasion joy and provide a refuge in which sorrow can speak and be heard, in which the meaning of our lives can be honestly expressed and considered. Good relationships sustain us amidst the frustrations and troubles we face across our lives. Theologically speaking, such relationships—"right relationships"—incarnate grace. Biblical images for right relationship are surprisingly few and precious: David and Jonathan, Ruth and Naomi, Mary and Elizabeth, Jesus and the centurion, Jesus and the woman who anointed his feet, the injured man and the Samaritan. Theologians today often speak of right relationship as "authentic," "mutual," "real," or "liberating" relationship, by which they mean to convey the experience of being in relationships that make us feel more alive, more connected to God, more aware of and thankful for who we are, and more sensitive to those around us and our surroundings.

Of course, not all the relations we find ourselves in are liberating or mutual. Some we mistakenly embrace, others we tolerate or avoid. Some we unconsciously accept, others we consciously endure. At their worst, harmful relations rob us of joy and security. They undermine our sense of belonging and hope. They strike at our weaknesses and entrap us by obscuring the way out. They engulf us in destructive visions of life, which, although perhaps illusory at the outset, can become all too real and injurious as time goes on. Sometimes harmful relationships can even change us, without our knowing it, into people we never wanted to be. Theologically speaking, such relationships— "broken relationships"—are highly complex manifestations of the pain that flows from the reality of human sin. The Bible abounds with images of broken relationship: Cain and Abel; Sarah and Hagar; Joseph and his brothers; Pharaoh and the Israelites; David, Uriah, and Bathsheba; Samson and Delilah; Jesus and Judas. Theologians often speak of broken relationship as the experience of being driven by our surroundings into self-loathing or self-adulation such that we fail to sustain ourselves and our connections to God, others, and the world. Of course, most biblical relationships are neither simply good nor bad but a complex combination of both, embodying elements of grace and brokenness together: Adam and Eve, the Israelites and Mt. Sinai, Jonah and the whale, Daniel and the lion's den, Jesus and Peter, Paul and the churches he visited.

The final chapter draws on the relationality of quantum entanglement to name afresh, and in a way that connects us more broadly to the world, both "right" and "broken" relationship. Chapters 3, 4, and 5 explore just what type of relationality entanglement is. They also aim to give the theological reader a better sense of how the physics and the physicist's mind actually work. For the remainder of this chapter we will stay on a theological trajectory, examining some of the ways in which the category of "relationality" and the modifiers "right" and "broken" have come to function in contemporary theological writing. Although theological deployments of relationality can be a bit fuzzy at times, there is much to be gained by thinking theologically about, and with, this powerful idea. Right relation, as I will use the term in the following—on the basis of my own communally formed interpretation of the biblical witness and relational view of God—connotes mutuality of freedom and responsibility, equality of power, reciprocity, intimacy, honor, fidelity, responsiveness, participation, empathy, understanding, communion, reconciliation, and fellowship.

4 MULTIPLE POINTS OF ENTRY

A helpful way to continue the narrative at this point is to consider a sampling of recent theological literature that invokes the category of relationality. The overviews and evaluations that follow are intended especially for those who are unfamiliar with the literature, but they also function collectively both to substantiate my claim that a "turn toward relationality" has indeed taken place in recent theology and to indicate the wide variety of topics and perspectives that now span the discussion. I have organized these discussions by theological theme in order to highlight the breadth of topics present in the literature. This arrangement should not be mistaken for an attempt to cobble together a more comprehensive theological vision around the idea of relationality—the diversity of viewpoints among the writers is too great (I will gesture toward a larger vision in the final chapter). Nor do I take the time to critically analyze each of the writings in detail, which would require another (and very different) book. What follows can be understood as a sympathetic exposition and evaluation of some of the more prominent contributors to the recent theological literature on relationality. I should also mention that a significant number of the theologians discussed in the following either work from or are strongly influenced by a process perspective. As I mentioned earlier, many of those charting out new theological territory on the subject of relationality have been process thinkers. Additionally, a good number see themselves as contributing to the revival of trinitarian thought. I have connected each theologian to a particular theme, but it should be apparent that nearly everyone discussed could have been associated with others as well. The writings span the themes of God, anthropology, Christology, soteriology, pneumatology, ecclesiology, ministry, politics, the world's religions, and eschatology. I leave creation to the next section, where it receives extended treatment.

God

It is perhaps fair to say that the bulk of recent theological writing on relationality has focused on the idea of *divine* relationality. One recent example is Karen Baker-Fletcher's reappraisal and reconstruction of the trinitarian tradition in her *Dancing with God: The Trinity from a Womanist Perspective* (2006). Taking womanist theology to be inherently integrative and relational, Baker-Fletcher argues that God is present in the world, healing and "whole-making" the lives of those affected by violence (2006, ix). To this end she construes the three-ness of God perichoretically as a dance or "community of movements" (45). Appealing both to the Cappadocians (55–56) and to process thought, she blends the two traditions by redefining omniscience and omnipotence with an eye toward relationality, rather than giving up such categories altogether. Baker-Fletcher's relational God dances with creation, courageously and graciously responding to unnecessary violence in the form of Jesus's Spirit-powered ministry. Followers of Christ can learn to partner with God in this dance. Baker-Fletcher draws on the stories of Alice Walker, W.E.B. Du Bois, James Cameron, James Byrd Jr., and many others to weave an intricate and thickly personal narrative of prophets, victims, poets, and visionaries who have danced with God. She is both appreciative and critical of the trinitarian tradition as she takes hold of ancient concepts and attempts to breathe them into new life. Baker-Fletcher does well what many of the recent discussions on divine relationality do—she creatively brings fresh insight to traditional ideas and effectively intertwines social awareness with spiritual yearning.

Anthropology

The emphasis on divine relationality has its anthropological counterpart in the literature as well. In *Many Voices: Pastoral Psychotherapy in Relational and Theological Perspective* (2007), Pamela Cooper-White characterizes the Trinity as a "fluid *metaphor* that challenges . . . singular, totalizing images of God" (2007, 79; italics in original), a theological resource for reconstructing theological anthropology around the notion that human beings are not merely relational but also *multiple*. We are multiple, she argues, in the sense that we have diverse relations with one another and because of the internal multiplicity that characterizes our own personhood. Cooper-White wields postmodern accounts of plurality, postcolonial notions of hybridity, and contemporary relational psychoanalysis to construct a model of pastoral psychotherapy that "seeks to help individuals come to know, accept, and even appreciate all the distinctive parts—the many voices—that live within them" (viii). She constructs her theological anthropology around the core idea that human beings are variegated rather than single, unified selves, paralleling and grounding this conception of human personhood in the multiplicity of the trinitarian God. As a relationally minded pastoral psychotherapist, Cooper-White takes her task to be one of helping people discover new capacities for relationality, both external and internal, over and against the potential loss of self-integration that comes with, for example, the externalization ("splitting") of threatening

aspects of one's inner life. Her provocative and dense blend of psychothera-
peutic and trinitarian perspectives yields a distinctive anthropology that
values multiplicity even as it works against fragmentation. Cooper-White's
work calls attention to the delicate balance that must be struck by any theological account of relationality between the generativity of multiplicity and
the equanimity of wholeness.

Christology

Also in the psychological vein, Charlene P.E. Burns's *Divine Becoming:
Rethinking Jesus and Incarnation* (2002) develops the claim that theologians too often privilege philosophical categories over psychological
ones in their accounts of the nature of Jesus Christ. Burns constructs her
Christology on the basis of psychological and anthropological accounts
of human development, in particular discussions of entrainment (rhyth-
mic physiological synchrony), attunement (awareness of others' emotions),
sympathy (feeling for others), and empathy (feeling with others), as found
in the works of anthropologists Edward T. Hall and Max Scheler, as well
as psychoanalytic theorist Daniel Stern. She groups these ideas under the
general heading of "participation," which has a long pedigree in Christian
thought around naming the creaturely experience of the trinitarian God.
Jesus as the Christ, in Burns's view, "manifests the epitome of relatedness:
he entrains with, attunes to, feels sympathy [for] and empathy with humanity and God in such a way as to be a singular revelation of the divine"
(2002, 17); he uniquely and fully incarnates the empathetic participation
of God in creation. According to Burns, however, *all* of creation has the
potential to incarnate the divine in this way. She develops her wide account
of incarnation in the direction of a "radically communal vision of life . . .
[in which] individuality is preserved as constituted within the divine matrix
of relationality" (158). The fullest expression of "participatory relation" is
God's own self, which God knows and experiences "as a unity gone out to
fullest self-presence in the plurality of being" (136). Burns's wide Christology aims to preserve the uniqueness of Jesus, but also to connect God's
being to the idea of relationality in such a way as to enable all creation to
participate in the divine life. There is much in her account that resonates
with the view I develop in Chapter 6, especially her notion of a communal
life that sustains individuality.

Soteriology

Broken and reordered relationships lie at the heart of David L. Wheeler's
soteriological project, *A Relational View of the Atonement: Prolegomenon
to a Reconstruction of the Doctrine* (1989). Like other relational thinkers
influenced by process thought, Wheeler is motivated to rethink the meaning of soteriology through the category of relationality on grounds that "all

occasions of experience, including human beings and God, are constituted as the existents they are in and through their relationships" (1989, 190). This leads him to understand atonement as a process in which both creation and God are enriched through their relation to one another. From the human perspective, Wheeler contends, we actually become something new because of what God does rather than simply being accorded a new status. Whereas broken relationships harm us in our self-constitution, the soteriological reordering of these relationships positively reconstitutes us as selves. One can hear in this assertion an echo of the process doctrine of internal relations, i.e., the notion that entities are what they are by virtue of their relations to other entities (Whitehead 1967). In Wheeler's account, Christ becomes the image and supreme instance of the unity of divine love and creaturely response. Like Burns, Wheeler aims for a wide view of this unity that does not relinquish the historical uniqueness of the person of Jesus Christ. His appeal to process categories provides him a way of asserting God's involvement in creation without denying the full freedom of creation to become itself. In my own account, I go after this problem from a different direction because of the way in which I construe entanglement for theology. God's entangled relationship with the world is a relationship that grants it full freedom. God, as relationality, desires to be in right relationship and nothing more—creation is truly free to be itself.

Pneumatology

Recent relational work in the area of pneumatology takes Augustine's view of the Holy Spirit as the bond between the Father and Son (1963, Bk. 6.5) a step further. F. LeRon Shults and Steven J. Sandage, in *Transforming Spirituality: Integrating Theology and Psychology* (2006), for example, develop a pneumatology that highlights the spiritual transformation they regard as necessary for living a meaningful life. Can the Holy Spirit be understood as the source of our longings for intimate relation with others and with God? For Shults and Sandage, the Spirit is the very principle of relationality within the relational triune God, awakening and arousing us, "calling us into intimate fellowship with God in Christ" (2006, 41). Their relational pneumatology involves three moments of retrieval: first, they reconfigure the concept of divine infinity so as to undermine any sharp distinction between God as one type of substance and creation as another; second, they rework the Eastern notion of *theosis* in terms of entering into the divine life rather than becoming part of the divine substance; and third, they conceptualize an eschatological ontology in which the "temporal existence of creaturely desire [is] constituted by the promising presence of the Creator Spirit, who graciously opens time into eternity" (55). For Shults and Sandage, a human life lived *in the Spirit* is a life of knowing, acting, and being oriented toward Infinity, Trinity, and Futurity. By emphasizing the communal and bodily nature of human experience, they intend to guard against the individualist and isolationist tendencies of early modern anthropology. The human desire to "belong-to and be longed-for . . . [is the arena within which] a person's feeling of and for freedom of being-in-relation emerges" (123). The divine Spirit mediates this emergence, not as one being among many but as the "all-embracing presence in which we already 'live and move and have our being' (Acts 17:28, NRSV)" (138). To be transformed by God, according to Shults and Sandage, is to be welcomed by the Spirit into the perichoretic relationality that is the inner life of the trinitarian God. Most pertinent to my own desire to rethink the *creatio ex nihilo* tradition in more explicitly trinitarian terms as *creatio ex relatione* is their attempt to work beyond the dualism of God as one type of substance and creation as another. As I will argue in Chapter 6, a trinitarian account of creation should avoid saying that creation is not God, just as much as it should avoid saying that creation is God. If creation is not "brought" but "related" into existence by a relational God, then the relationship between God and creation must transcend the difference between absolute identity and absolute difference.

Ecclesiology

In *After Our Likeness: The Church as the Image of the Trinity* (1998), Miroslav Volf advocates a Protestant "free church" ecclesiology over and against the Roman Catholic ecclesiology of Joseph Ratzinger (now Pope Benedict XVI), on the one hand, and over and against the Eastern Orthodox ecclesiology of John Zizioulas, on the other. Volf's primary interest is in the relation between person and community vis-à-vis the church and its members. For Volf, "If Christian initiation [baptism] is a trinitarian event, then the church must speak of the Trinity as its determining reality" (1998, 195). Wanting to fend off both overly individualistic (read: Protestant, free church) and hierarchically holistic (read: Roman Catholic and Eastern Orthodox) accounts of how the church images the Trinity, Volf deploys a perichoretic, nonhierarchical interpretation of the Trinity and argues that the presence of Christ in the church is mediated through the *whole* of its members; in local terms, this means through the *entire* congregation. It is wrong, Volf thinks, to say that God acts externally in creation as one tripersonal self; one must instead say that God acts "only as a communion of the different persons existing within one another" (215). The church, Volf thinks, should pattern itself and its own action after this relational, egalitarian understanding of the divine life. Persons who are part of an ecclesial community have the right to be heard and participate in its governance. They have these rights because, as centers of action, they "correspond to the relations of the divine persons," but equally important, they *must* have rights because they and the church still live "on this side of God's new creation," where the potential for abuse is an ongoing concern (220). The vivid egalitarian hues of Volf's relational ecclesiology and Trinity reflect the broader concern among those writing relational theologies that the mutuality of the trinitarian persons might find full expression in human life, especially in the life of the church.

Ministry

Coming from a pastoral perspective (very different than that of Cooper-White), Graham Buxton pleads for those engaged in pastoral ministry to be more attuned to the impact of science and technology in the lives of their communities and to the struggles of those who attempt to negotiate the complexities of being scientists *and* religious practitioners. Buxton's *The Trinity, Creation, and Pastoral Ministry: Imaging a Perichoretic God* (2005) forges a pastoral theology that brings together trinitarian thought, biology, and contemporary physics (including entanglement, which I will return to in Chapter 6). For Buxton, the pastoral potency of *perichoresis* follows from the power of the relationship made possible between God and creation when creation participates in the divine life. Emphasizing diversity-within-unity, Buxton argues that pastoral practice must embrace the particularity of specific lives and specific cultural, psychological, economic, and social contexts. Those engaged in Christian ministry must be mindful of the webs of mutuality and participation that have both anthropological and cosmological ramifications. Noting that the trinitarian God is not just relationality but *communal* relationality, Buxton considers the nature of Christian ministry under several "perichoretic themes" (2005, 149–193): community formation (difference, diversity, polyphony), community realization (agapistic love, hospitality, *ekstatic* orientation), and finally, community operation (worship, mission, compassion). He elaborates an account of pastoral care and compassion around the idea mutual giving and receiving, which he suggests "resonates effectively with the trinitarian experience of *perichoresis*" (193). The idea of mutual indwelling, giving, and receiving is attractive to Buxton because it provides him with language for the divine that can function to name "right relation" between the individual and the community—between the one and the many on any number of theological issues—thereby simultaneously maintaining individual identity and mutual interdependence. Once again we have a relational theologian working to strike the right balance between these two poles. One reason to be attracted to (and cautious about) entanglement as a theological metaphor is that it contains within itself a particular view of this balance—entanglement is not just generic relationality, but a highly specific form that will push theological discussions in particular directions, as we shall see.

Politics

How shall we live our lives together, and what roles should we play as persons within a larger society? Douglas Sturm probes these two fundamental

questions at the intersection of politics and religion in his book *Solidarity and Suffering: Toward a Politics of Relationality* (1998). According to
Sturm, the primary concern of the public forum is to "understand ourselves,
our relationships with each other, our place in the world, our responsibility
to the future, our participation in the whole ongoing community of life"
(1998, 3). We are, he contends, presently in the midst of a crisis of global
proportion that threatens the health and perhaps even the survival of life
on Earth. The crisis has social, ecological, economic, and political dimensions, none of which can be addressed in isolation from the others. What is
needed, Sturm argues, is a new understanding of ourselves as members of a
"vast and variegated community of life" (4). Taking what he regards to be
the best of the politics of "welfare," the politics of "liberty," the politics of
"community," the politics of "difference," and the politics of "ecology," he
constructs a politics of relationality based upon his theological construal of
justice as solidarity. Communitarian theory, democratic socialism, religious
pluralism, nonviolence, and deep ecology all contribute to this broadly conceived and masterfully executed politics of relationality. Resisting Reinhold
Niebuhr's distinction between love and justice (1964, 244ff.), Sturm applies
the Johannine notion of God as love to the political and personal realms
alike. His own style of theorizing, which he calls "koinonology" (1998,
275), is an excellent example of how the category of "relationality" can be
thickened and delimited in ways that give the term specificity and traction
in the service of theological interpretation.

The World's Religions

One might think that the particularity of trinitarianism would be a serious
impediment to constructing a theological argument for respecting the wisdom of the world's different religions. Mark Heim, in his *The Depth of the
Riches: A Trinitarian Theology of Religious Ends* (2001), turns this expectation on its head by arguing instead that a relationalist understanding of
the Trinity *demands* the truth of religions besides Christianity. Drawing
on Zizioulas's account of God as the communion of persons-in-relation,
Heim argues that the different dimensions of the divine life entail multiple
religious ends. Christian salvation is real, but so is Buddhist *nirvana*—
each religious end points in its own way to the complex communion-nature
of God. Just as the ontological difference between God and humanity is
not an obstacle but rather a means to salvation—the union of human and
divine in Jesus Christ being the Christian exemplar of this dynamic—so
too are differences between religions and religious ends not contradictory
but instead specific expressions of the communion of the divine life. "The
very fact that [human] being is constituted in relation with others, relation
with what is unlike . . . is the most fundamental way that we are like God"
(2001, 127). The Trinity, then, requires a pluralistic theology of religions,
or as Heim provocatively puts it, "It is impossible to believe in the Trinity

instead of the distinctive religious claims of all other religions" (167; italics in original). Instead of leading to exclusivity, here the particularity of a God who is intrinsically relational motivates an open posture to the truths contained within the world's many religions.

Eschatology

In the West we are (let us hope) becoming more keenly aware of the fate of the many, the fate of the whole—especially through the work of environmentalists, the threat of nuclear annihilation, the so-called "flattening" of the world (Friedman 2006), and even advances in scientific cosmology that bring more and more of the universe within intellectual grasp. As a result, the doctrine of the "last things" need no longer be reduced to a theological curio or Christian-speak for social progress. Rather, it is increasingly regarded as an indispensible theological category for orienting Christian life and theology, especially in relation to the idea that God's saving work in the world is bringing about the consummation and perfection of the entirety of creation (see, for example, Moltmann 1993c; Peters 2000; Polkinghorne and Welker 2000; Peters, Russell, and Welker 2002; Ellis 2002; Barber and Neville 2005; Russell 2006; Wright 2008). Personal eschatology is also being rethought from a relationalist point of view. Anthony Godzieba, in his essay entitled "Bodies and Persons, Resurrected and Postmodern: Towards a Relational Eschatology" (2003), considers the meaning of eschatological language in the particular context of the loss of a loved one. Godzieba rejects a dualist anthropology, which had the merit of allowing mourners to "make sense" of death as something less than a complete loss, but he wants to keep "soul" language for naming that which constitutes us as whole persons and which includes but goes beyond our bodiliness. Instead of giving up on "soul" language, he thinks we should free it from the chains of dualism so that it can once again function effectively as a symbol that expresses "our limitless love and our devastating loss" (2003, 224). Godzieba draws on the philosophical anthropologies of Hans-Georg Gadamer and Manfred Frank to argue that soul language, and eschatological language in general, cannot function in our own time unless it is grounded in an embodied, relational notion of personhood. He admits that a strongly realist view of the resurrection lies beyond what rational thought can achieve but thinks nonetheless that relationalist anthropologies such as those of Gadamer and Frank can help theologians resist a pernicious "reduction to biology" when thinking about personhood. My own view of eschatology, which I lay out briefly in Chapter 6, contrasts sharply with these attempts to justify the consummation of all creation on the basis of a God who transforms the world by fiat. I prefer to speak of eschatology in temporal terms as the actual possibility of a future, even a better one, entailed by God's gracious will that creation simply be.

This sampling of recent theological literature on relationality (see the Appendix for a more comprehensive list) points to many of the different interests that have surfaced around the concept of relationality over the last several decades. One general and important limitation of this literature is its bias toward using the language of divine relationality exclusively for the purpose of contextualizing the meaning of *human* relationality. In the next section, I take up the thought of two theologians who have appealed to the category of relationality in conjunction with a broader set of concerns having to do with the idea of creation. Interacting with these theologies will move us in the direction of the final chapter, where "entanglement" becomes an overarching image for the relationality of God, of creation, and between God and creation.

5 TWO RELATIONAL VIEWS OF CREATION

Christian theologians of the second and third centuries contended that the world is neither eternal nor evil. They understood the world to be created good by God, as Genesis so eloquently proclaimed, and moving toward ultimate perfection and participation in God. The world's radical dependence on God—the contingency of creation and the freedom of the divine creative act—came to classic formulation in early Christian thought in the doctrine of *creatio ex nihilo*, which first appeared in the second century C.E., prior to the development of the doctrine of the Trinity, in the writings of Tatian, Theophilus of Antioch, and Tertullian who argued against Platonic and Gnostic readings of the Christian story. As Colin Gunton pointed out (1998), however, alongside this undifferentiated version of *creatio ex nihilo* there developed a proto-trinitarian perspective of divine creative activity. Irenaeus of Lyons, for example, affirmed *creatio ex nihilo* but also taught that God the Father created the world with two divine hands: the Son and Spirit (1992, 4.20.1). Gunton argued for a recovery of a trinitarian view of creation after what he called the "Babylonian captivity" of the doctrine, i.e., a nontrinitarian, Neoplatonic emanationism that took hold during the Middle Ages.

A response to Gunton's concern appears in the theological project of Wolfhart Pannenberg, whose larger corpus has been shaped by a desire to bring theology out of its "Barthian ghetto" and into conversation with other forms of knowledge. While embracing much of Barth's theological perspective, particularly his insistence on the trinitarian nature of God as the all-determining subject of Christian theology, Pannenberg has resolutely rejected Barth's self-imposed isolation from modern science and philosophy. He has developed instead a theology that aims to recognize the structures and limits of human knowledge even as it maintains a vibrant evangelical and future-oriented posture. Here I will focus on Pannenberg's discussion of creation in volume II of his *Systematic Theology* (1994).

Pannenberg's view of the world as creation is rooted in his understanding of the Trinity. The goal of creation, he argues, is the "participation of creatures in the Trinitarian fellowship of the Son with the Father" (1994, 73).[2] He affirms the absolute freedom of God in the act of creation but wonders whether this act has "a share in the reference of the divine persons to one another that leaves them indivisible" (4). Pannenberg rightly denies that God needs the world in order to be active and, viewing creation as an act of love, argues instead that the "freedom of the divine origin of the world on the one hand and God's holding fast to his creation on the other belong together" (19). Pannenberg's interest in sorting out the relation between the freedom and necessity of the divine act of creation raises a related question regarding creation itself—a question that goes to the heart of the present work. Does creation *itself* as the product of an external (to God) act share in any way in the reference of the trinitarian persons to one another that leaves them indivisible, notwithstanding the world's obvious multiplicity? In other words, does the claim that the world was created by a trinitarian God entail that creation is a "structured and differentiated unity" whose unity "is not lost by reason of the plurality of events" in creation (1994, 8, 9; cf. 34–35)? For Pannenberg, the world has its unity-in-plurality because it is the creation of a God whose very being is unity-in-plurality.

But if creation's relationality has its basis in the divine life, then what distinguishes God's external acts from God's internal acts? The difference, for Pannenberg, lies simply in their outward character (4, 5). The mutual relations among the trinitarian persons constitute the self-activity of God; the trinitarian God does not need the world to be active. The creation of the world, on the other hand, involves the trinitarian persons moving together out of the divine essence. But if one is to understand creation in light of the internal activity of God, which is the basis of God's external acts, then one must not draw too sharp a line between internal and external divine acts. The latter are nothing more (and nothing less!) than the internal acts of God externalizing themselves. The divine act of creating a world implies for Pannenberg not only the world's existence but also its redemption and consummation (7). In this way, Pannenberg sees creation-salvation as a unified, supratemporal divine act; the entire unfolding of creation *is* God's creative act from the perspective of divine supratemporal eternity.

Curiously, Pannenberg pursues the issue of creation's unity only with regard to temporality. A temporal sequence of events may still be thought of as a unity, he argues, in light of the integrating effect brought about by the end or purpose of the entire sequence of events. Singleness of purpose is what allows us to speak of a temporal series of events as a single whole; temporality is related to eternity "in terms of its longing for wholeness and identity" (1988, 326). But what about creation's spacial status from moment to moment? If Einstein's view of the relationship between space and time is correct, one cannot picture the two as being entirely separate (more on this later). Do the different spatial parts of creation admit any type of wholeness beyond that of being temporally bound together by a divinely ordered end? Pannenberg fails to consider the question of creation's unity from the perspective of space (or better, space-time). At the end of Chapter 5 and again in Chapter 6 I will argue that quantum entanglement can be understood as a trans-spatiotemporal phenomenon that allows for an even more robust theological conception of the unity of creation.

For Pannenberg, the Godhead is characterized by ontological reciprocity. The Father is the Father precisely because of the loving relation between the Father and the Son. Without the Son, the Father would not exist *as* the Father who loves the Son. The love of the Father also sustains creation in its existence and independence. Does this imply the divided attention of the Father? No, for the loving goodness of the Father, which is the origin of all things, is no different from the love which the Father directs toward the Son. This is so because the Son is the principle of creation's existence and particularity. The self-distinction of the Son from the Father "forms a starting point for the otherness and independence of creaturely existence" (1994, 21). The significance of the incarnation is not only soteriological for Pannenberg but cosmological as well (cf. Rahner 1976, 179). The New Testament builds on the Hebrew Bible's account of the cosmic role of divine Wisdom (Prov. 8:22–31), although it expresses this idea primarily through the Greek concept of *Logos* (John 1:1ff., Col. 1:15–20, Heb. 1:2ff.).[3] The Son, then, is creation's formal principle, the basis of the possibility of its existence "outside" of God, and at the same time its substantive principle, i.e., the Son's actual turning outside the Godhead *is* the act of creation from the divine perspective. The very possibility of existence "outside" God has its basis in the second person of the Trinity as the emblem of the inner-differentiated nature of divine being, and the actuality of creation has its basis in the specific act of second person turning the Trinity outward (i.e., in the incarnation)—recall the traditional distinction between the *ad intra* and *ad extra* acts of God. Pannenberg distances himself from Hegel on this point, arguing that the mutuality of the relations among the trinitarian persons make for a self-completed circle that does not need to create (to posit the other) in order to attain self-knowledge. The outward turn of the second person is more than a movement necessitated by the logic of distinction for Pannenberg; it is a free act (1994, 30).

Pannenberg notes that early Protestant theologians, in particular those with a pietist bent, confined the role of the Spirit to matters of soteriology: "the Spirit became a factor in subjective experience rather than a principle in [the] explanation of nature" (1994, 127). German idealism recovered the creative role of the spirit under the influence of Hegel, but there the equating of spirit with mind led to Feuerbach's charge of projection. This historical backdrop motivates Pannenberg to construct a pneumatology that reaches behind what he sees as the "subjectivist thrust" of the doctrine in recent centuries to uncover the active role of the Spirit in all creation. The Spirit, he argues, binds the Father and the Son in the unity of their free decision, which Pannenberg identifies as an expression of fellowship (30). The work of the Spirit is closely related to the work of the Son, although different in character. Whereas the Son is the principle of distinction within and outside the Godhead, the Spirit is the corresponding principle of fellowship and participation. Pannenberg calls the finitude-transcending participation of creation in the divine life "the special work of the Spirit" (33). He boldly identifies the relational dynamic of life as the presence of self-transcendence in creation. Unfortunately, his account of the work of the Spirit leaves out most of creation—the inanimate world—putting it at some conceptual distance from God's creative, relational work.

For the past several decades Sallie McFague has been calling North Americans to take seriously issues of environmental and economic justice. According to McFague, Christian theology's central task of expanding models conceptually to provide a more coherent and comprehensive account of belief depends crucially upon its ability to wield the "root metaphor" of the "kingdom of God." Referred to in Jesus's parables and pointed to by Jesus himself, the kingdom of God points to relationality as the fundamental theme of Christian theology (McFague 1982, 108). The kingdom of God disorients away from a life of merit and toward a life of grace, a life fully aware of its own relationality. And precisely because McFague sees the notion of relationality, and not something else, as the meaning of Christianity's root metaphor, she insists that theology must be metaphorical in both content and method. If the tension of metaphorical language is lost, then the dynamism of the concept of relationality will be lost as well. Blending insights from process philosophy, feminist epistemology, and ecology, McFague explores multiple models of self and world that affirm the importance of interconnectedness while at the same time allowing the other to be the other (1997, 98).

For a theological expression of the Christian message to be relevant in our own time, McFague argues, it must focus both on the rise of the dispossessed in terms of gender, race, and class, and on the growing awareness of the interdependence of life at all levels (1982, xi). Prior to the modern period, humans saw the world through a sacramental lens. The medieval codification of Aristotle's "great chain of being" led to the understanding that "God ties everything together in a silent ontological web which reverberates with similarity within dissimilarity out to its farthest reaches. . . . In such a universe, everything holds together, everything fits, everything is related" (1982, 6). McFague wants a revitalization of religious language that can recapture the wholeness of the sacramental view, but she thinks we cannot simply return to premodern sacramentalism. Instead, we need to draw upon current ecological perspectives to resharpen our ability to perceive the world through synthetic/holistic rather than analytic/reductionistic categories. According to McFague, such an ecological or organicist worldview looks to the whole first, focusing on the interrelations and interdependence of the parts, whereas the mechanistic view puts parts first and regards them as the sole carriers of causal powers (1993, 15). She objects not to reductionism *per se* but to its refusal to grant the possibility that wholes can function as causes too. (When I use the word "reductionism" in the following, I have this exclusivist assertion in mind.)

McFague's well-known metaphorical description of the world as God's body, which receives fullest treatment in *The Body of God*, provides a clear example of her relationalist approach—all creation is the very flesh of God (1993, 133). Despite extending the notion of "God's body" to all materiality, however, she focuses almost exclusively on the realms of human and nonhuman life (97, 167). McFague rightly points out that personal metaphors are central to the history of Christian thought, which has been interested primarily in the relation between God and humanity as manifested in the person of Jesus Christ. But she also acknowledges that this bias toward personal metaphors should not keep us from exploring nonpersonal, naturalistic metaphors (1982, e.g., xi, 20, 167, 178). In fact, as McFague argues, the dominant place of personal metaphors within the history of Christian thought has been detrimental to the natural environment (1982, 20; cf. 178). Still, she thinks that personal metaphors must remain primary: first, because of their power to unseat the "triumphalist, royal" model, and second, because theological models must come from a place "deep within human experience" (1987, 80).

McFague's minimal development of naturalistic, inorganic metaphors—she names only a few (sun, water, sky, mountains; 1987, 82)—appears to stem from the judgment that naturalistic metaphors have little capacity to evoke the relationality of human experience (83). She is troubled by the fact that contemporary science does not allow for "any purpose or agency apart from local causation" (1993, 142). But as I will argue in the next three chapters, quantum entanglement presents us with a physical phenomenon in which local causation cannot be the whole story. McFague's strategy of countering the harmful ecological effects of the mechanistic worldview by promoting an ecological worldview has its own problem. An exclusively ecological or organicist worldview is, in its own way, as reductionistic as a mechanistic one. McFague develops her image of the world as God's "body," for example, in exclusively biological terms. She affirms that all materiality is effectively God's body, but she is content to take human and animal bodies as "representative" of the bodiliness of all creation (1993, 97, 167). In Chapter 6 I will argue that the physical phenomenon of entanglement clears a path toward the reintegration of the physical world into McFague's otherwise holistic, ecological approach.

McFague wants to view the world relationally, but whereas ecology proffers the idea that the world is composed of complex systems with many interrelated parts, it does not necessarily render "all talk of atomistic individualism indefensible" (1987, 8; cf. 11)—at least not from an ontological perspective. McFague's own account reflects this shortcoming insofar as hers is a "compositional" view of the world. However complex it may be, she takes the world to be constructed out of basic, independent parts. These parts have their existence, both logically and ontologically, prior to that of the whole. Although McFague initially turned to quantum physics to warrant the holism of her organicist model (1987, 8), she has subsequently relied exclusively on examples from the biological realm. Although her focus on ecology has yielded important fruit, she has overlooked the potential contribution contemporary physics can make to a relational worldview (10).

Can McFague's organicist approach help to "re-create" the world of rocks, protons, stars, and galaxies in the Christian theological imagination? Consider how McFague describes the three contexts in which theological reflection takes place: the psychological, political, and cosmological. She defines the first two contexts concretely as that of the individual and of human community, but she goes on to define the third context in terms of the "planetary" (2000, 30). This is not surprising, given how McFague's own approach is rooted in the distinctiveness of life on earth. In *Super, Natural Christians*, McFague asks whether nature is like us (1997, 46). Her point is to encourage us to develop a "loving" rather than "arrogant" eye toward nature, so that we can see nature as fellow subject and not merely as object to be gazed upon from the vantage point of our own needful subjectivity (1997, chap. 5, esp. 95–97). But the question is equally important when turned around: Are we like (the rest of) nature? Do our own impulses toward relationality and wholeness lie with the grain of the universe or against it? An appreciation of physical entanglement and its theological valence provides a way of reconnecting the physical world to McFague's ecological, relational vision.

Given the predominance of trinitarian thinking within contemporary relational theologies, it is interesting to see what McFague has had to say about the Trinity. Given that she is clearly no fan of the patriarchal aspects of the tradition, her assessment of the doctrine is surprisingly positive: "[The Trinity is a] conceptual attempt not to circumscribe the divine being in an absolute static formula . . . but to delineate the implications of the relational root-metaphor" (1982, 126; cf. 173–174, 190–191). The two central insights of trinitarian thought, according to McFague, are that God's being is characterized by relationality and that God has chosen to be in relation to something other than Godself: "The Trinity, then, far from being irrelevant, is central to Christian faith: it expresses the entire God–world dynamic" (2000, 143–144, esp. n. 17, where she approvingly cites LaCugna). Near the end of *Body of God* she writes that the "proper function of Trinitarianism in the Christian tradition is . . . to preserve for an agential theism both radical immanence and radical transcendence" (1993, 192) and advocates replacing the traditional language of "Father," "Son," and "Holy Spirit" with "Mystery," "Physicality," and "Mediation" (193). If we construe "Mystery" as ineffable, invisible source, "Physicality" as multiplicity and differentiation, and "Mediation" the relationality of divine being, a clear resonance emerges between McFague's formulation and Pannenberg's. In place of "Father" we find "Mystery," the unfathomable depth of divine being; in place of "Son" we find "Physicality," the possibility of creation latent within the inner-differentiation of the divine being; and in place of "Spirit" we find "Mediation," the relationality that reconciles otherness and unity.

For all their differences, Pannenberg and McFague are of like mind on the importance of divine relationality. Their ways of thinking about the world as the act of a relational God bring the doctrine of creation into closer contact with the doctrine of the Trinity than was initially possible in the early development of the *ex nihilo* tradition. Their approaches create the possibility of finding value in the world apart from its utility for human life. On the other hand, both Pannenberg and McFague follow the lead of the other relational theologians we have encountered by thinking of the "world" primarily in personal or perhaps biocentric terms, effectively rendering invisible that which would appear to comprise the vast majority of the universe: nonliving matter. The next three chapters address this problem by assembling the rudiments of a different mind-set, one that can appreciate the beauty of life without slighting the relational complexity of the physical world.

3 Separateness in Classical Physics

[According to quantum theory] one has to assume that the physically real in [location] B suffers a sudden change as a result of a measurement in [location] A. My instinct for physics bristles at this.
— Albert Einstein, in a 1948 letter to Max Born
(Born 1971, 164)

The previous chapter demonstrated that theologians from across the spectrum have lately been using the concept of "relationality" as a compass to orient their theological work. But theologians are not the only ones interested in the concept of relationality. Since the early part of the twentieth century physicists and philosophers have been having a debate of their own about whether a subtle type of relationality characterizes some, or perhaps even all, of the physical processes going on around us. In this chapter we consider the early part of that debate in several stages. We will begin by examining classical physicists' understanding of light in order to familiarize ourselves with three presuppositions that characterized their general understanding of physical processes through roughly the end of the nineteenth century, i.e., prior to the birth of quantum physics. I refer collectively to these three pre-quantum principles—property definiteness, state separability, and cause locality—as the "classical worldview." They are by no means the only principles that could be said to have characterized the mind-set of physicists prior to the twentieth century, but they are particularly relevant to the present project in two ways: They play a crucial role in the derivation of the Bell-type inequality presented later in the chapter, and they undergird my sense of the significance of quantum entanglement for theological discussions of relationality.[1] We then revisit the debate among the first quantum physicists—especially between Einstein and Niels Bohr—over whether or not quantum theory called the classical worldview into question. This debate subsided and lay dormant for several decades until the middle of the twentieth century, when physicist John Bell revisited Einstein's arguments and, much to physicists' surprise, proposed a crucial experiment. There had been no indication earlier that the differences between Einstein's and Bohr's views carried any experimental implications, but Bell showed that the basic principles behind Einstein's objections to quantum theory implied a specific outcome for a particular experiment. He also showed that this outcome was at odds with the quantum prediction. Beginning in the 1970s, physicists around the world embarked upon a quest to realize Bell's "thought experiment" in the laboratory. The broader physics community wanted to know whether or not the world followed classical rules, and, if it did not, whether the quantum prediction would be confirmed by what actually happens in a Bell experiment.

Before getting started, I want to pause for a moment to clarify an important point. It is potentially a source of serious confusion, and readers would do well to consider it before continuing: without a clear grasp of this point, one can easily misunderstand the spirit of what Bell was attempting to do. Here, then, is what one must keep in mind throughout the remainder of the chapter:

> The ideas and arguments presented here, including the Bell-type prediction derived at the end of the chapter, all follow from a *classical* perspective on physical processes.

In light of this fact, any discrepancy between laboratory results and the prediction that follows from our Bell-type inequality counts first as *a mark against the classical perspective*. In logical terms, the gist of Bell's prediction was simply "if A then B," where A is the classical worldview and B is a mathematical inequality related to a particular set of experimental results. If what actually happens in the laboratory is "not B," then the conclusion "not A" follows by *modus tollens*. In other words, a result "not B" tells us that something about the classical worldview "A" must be wrong. The possibility for confusion arises from the fact that Bell's argument has sometimes been misinterpreted as an argument based on quantum principles. This is not the case. Bell invoked no quantum principles in his derivation, and neither will I when I present my own derivation later in the chapter. The question of whether or not actual laboratory results confirm the quantum prediction is an entirely separate matter—in fact they do, but that is the subject of the next chapter. The subject of *this* chapter is the classical point of view and what it implies for how a Bell-type experiment ought to turn out. Along the way there will be occasion to indicate briefly where the "standard interpretation"[2] of quantum theory diverges from the classical perspective, but this will merely prime the pump for the next two chapters. So let us be clear. The primary task of this chapter is to introduce three key presuppositions underlying the *classical* perspective and then to follow in Bell's footsteps by examining what they imply for the outcome of a particular experiment.

1 INTRODUCTION AND BACKGROUND

We need to begin by coming to terms with why quantum theory initially seemed so counterintuitive to physicists who had been trained under the classical paradigm. The three presuppositions mentioned earlier bear directly on this question. Let me introduce them briefly here before examining each one in more detail and giving examples in the following section.

Property Definiteness

Classical physicists simply assumed—and they had no reason not to—that if some physical object has a property, it will necessarily have that property in a "determinate" or "definite" way. For example, if an object has the property of "length," this property necessarily has a single, well-defined value, say "4 meters." "Property definiteness" applies to every property an object has, and together the values of all of its properties comprise the object's "state." This brings us to the second hallmark.

State Separability

Classical physicists also assumed that an object's state[3] can be fully described without reference to the state of any other object. This is not to say that objects cannot affect one another, or that their states do not sometimes depend on their relationships with other objects. What "state separability" means is that one can fully describe the state of a physical object as something that belongs to that object alone, separately from the states of other objects. From a classical perspective, objects may influence each other's states but they do not *share* states—each object has its own.

Cause Locality

Finally, classical physicists assumed that the causal interaction of distinct objects is always mediated by a series of "local" events. Objects that do not touch one another can influence one another only indirectly, through the matter and energy that occupy the space and time between them.[4] The specific qualities of whatever matter or energy mediates an interaction determine how fast the influences propagate. Consider the act of "having a conversation." The fact that two people can easily converse across the street is possible because of the properties of the air molecules between them, i.e., because the speed at which disturbances propagate from air molecule to air molecule is quite fast, on average about 340 meters per second at sea level. When one person speaks, she sets air molecules in motion around her mouth. These molecules hit neighboring molecules, which hit their neighbors until finally the pressure fluctuations produced around her mouth by her voice propagate through local cause and effect across the street to the other person's ear. There is nothing causally "nonlocal," i.e., instantaneous or unmediated, about such an interaction, despite the fact that these two people are standing on opposite sides of the street. According to the classical principle of cause locality, all physical causal interactions are like talking in this regard; they are mediated by causal chains whose links are localized objects and processes that influence their immediate neighbors, who influence their immediate neighbors, and so on, at some finite speed.

In order to consider more closely what these presuppositions mean in the following sections, we will enlist the help of light. In particular, we need to consider its particulate nature, its polarization, and its speed. The most difficult part of this task is that we must try to adopt the mind-set of an early twentieth-century physicist. At that time, before the full development of quantum theory, physicists knew about the property of light called "polarization"—which pointed to the wave nature of light—as well as Einstein's 1905 argument for the existence of light particles, which he dubbed "photons." They did not yet have the benefit of a complete quantum framework within which to reconcile light's wave-like and particle-like behaviors. Our task for the remainder of this chapter is to think "classically" about the polarization of photons. The ideas put forward will help us to better understand the principles of property definiteness, state separability, and cause locality, and will eventually allow us to construct a Bell-style argument that leads to the "classical" prediction.

Photons are relative latecomers to the conceptual world of modern physics. Isaac Newton famously conjectured that light was composed of particles—he called them "corpuscles." What little he knew about the polarization of light led him to suggest that light corpuscles must be asymmetrically shaped in some way that determines whether they pass through a transparent material unhindered or are deflected from their ordinary path. (The so-called, slightly displaced "ordinary and "extraordinary" images obtained by looking through a birefringent calcite crystal were well known in Newton's time.) His corpuscular view held sway for over a century but fell from favor in the early nineteenth century when Thomas Young, Augustin Fresnel, and others demonstrated that light behaves like any other wave, i.e., it exhibits refraction, diffraction, and interference.[5] The final blow to Newton's particle theory appeared to come with James Clerk Maxwell's wave-theory of electromagnetism in 1864—one of the greatest achievements of nineteenth-century physics—which elegantly portrayed light as interwoven, undulating electric and magnetic fields. Imagine physicists' surprise, then, at Einstein's 1905 discovery that some important outstanding problems in basic physics could be resolved by assuming that light does indeed have a particulate (corpuscular) nature. More recent developments in quantum theory have put photons on firmer theoretical ground, leading to the notion that light particles are indeed instances of the indivisible "quantum" of electromagnetic energy. Prior to the appearance of quantum theory, however, physicists were left to ponder the apparently contradictory evidence on the nature of light from an entirely classical perspective. Today, a century later, physicists commonly speak of light as being composed of particles that exhibit wave-like characteristics.[6] In fact, this is how they talk about quantum particles generally (protons, electrons, etc.).

2   PROPERTY DEFINITENESS, STATE SEPARABILITY, AND CAUSE LOCALITY

Already in the early 1800s physicists understood that light manifests a kind of directionality called "polarization," which is orthogonal to its direction of propagation. According to Maxwell's theory of electromagnetism, the polarization of a beam of light corresponds to the direction in which its electric field undulates. Most light beams are "unpolarized," not in the sense of having no polarization but in the sense of having equal amounts of different polarizations. The simplest case of polarization—commonly

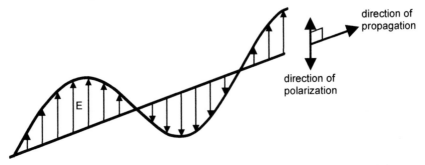

Figure 3.1   Polarization of light. The electric field E of a light beam undulates in some particular direction, called the beam's "polarization." The direction of this wave-like undulation is always "transverse" (i.e., perpendicular) to the direction of the beam's propagation. Simple "linear" polarization is shown here in the vertical direction, though it could have been shown along any transverse direction.

referred to as "linear" polarization—is one in which the beam's electric field undulates in one particular direction (see Figure 3.1).

Consider what happens when a beam of vertically polarized light encounters a "polarization filter" (see Figure 3.2), which has a rotatable "axis of transmission." As one might suspect, when such a beam encounters a filter whose axis of transmission perfectly matches its own vertical polarization, the beam passes through the filter unimpeded and arrives in full strength at a detector positioned on the far side of the filter—as though it were a wave crashing unimpeded on the shore. When the beam's polarization and the filter's transmission axis are "perfectly misaligned" or "orthogonal" to one another (i.e., the angle between them is 90°), the beam is completely blocked—as if it were a wave meeting a solid breakwater. When the beam and filter are misaligned by any other angle, some of the beam passes and some is blocked—as if it had encountered a breakwater with gaps. Although not perfect, the breakwater analogy provides some indication of how physicists thought about polarization within the "wave" account of light prior to Einstein's "photon" idea.

Can we explain the effect of the filter on the beam from the particulate perspective, i.e., from the perspective of photons? In order to do so, we must revert to something like Newton's conjecture that corpuscles are asymmetrically shaped in such a way as to make their successful passage through a medium depend on the relative "fit" between the shape of the particles and the composition of the medium. We will imagine the beam of light, therefore, as a collection of photons all traveling in the same direction and all having the same orientation in space, i.e., the same polarization. When the filter's orientation "matches" the photons' polarization, every photon is transmitted. When the filter's orientation

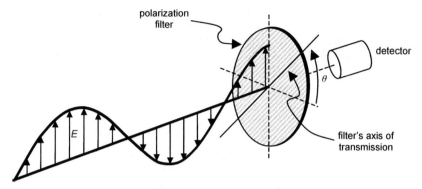

Figure 3.2 A vertically polarized beam of light encounters a polarization filter whose axis of transmission is oriented at angle θ with respect to the positive horizontal axis. If the filter's transmission-axis angle matches that of the beam's polarization (θ = 90°), the detector will register the beam at full strength. If the two are misaligned by 90° (θ = 0°), the detector will register nothing. If they are misaligned by any intermediate amount (0° < θ < 90°) as they are here, the detector will register the beam at some reduced strength.

is "misaligned" from the photons' polarization by 90°, every photon is blocked. This much we can make sense of: The fit is perfect at 0° and as poor as possible at 90°. But what about when the filter's polarization and the photons' polarization are "misaligned" by some angle other than 90°? As misalignment increases (going from 0° to 90°), fewer and fewer photons reach the detector. Furthermore, there seems to be no pattern as to which photons get through and which photons don't. Given that every photon started out with exactly the same polarization (i.e., orientation), this is an extremely puzzling result.

As a first step toward a working classical model of photon polarization that does not rely on a wave-conception of polarization (the direction of undulation), we must consider what we can and cannot learn about a single photon's encounter with a polarization filter. If an arbitrary photon passes through an arbitrarily oriented filter, we cannot know for certain what the initial polarization of the photon was because we know that photons are sometimes transmitted even when their polarization and the filter's transmission axis are somewhat misaligned. Furthermore, we cannot reorient the filter to perform a second measurement because the photon is inevitably destroyed in the process of the first measurement, either by the filter or the detector (we will return to this important point later). The only thing we *can* say about the photon is that it was in fact transmitted or stopped by a filter oriented in some particular direction. What we want is a way of describing photon polarization from a particle perspective that is consistent with the limitations pertaining to what we can measure.

Consider the following concrete example. A collection of 90°-polarized photons is incident on a filter whose orientation (i.e., transmission axis) is also set to 90°. A detector placed behind the filter will register "Transmit" (T) for every photon. This result does not tell us anything new, but it does present us with the opportunity to reexpress the photons' polarization in a way that is more specifically related to the actual measurement we perform and more amenable to a particle interpretation. Instead of saying that the photons were initially "90°-polarized," let us say that they carried the "T" value for the property "90°-polarization." For short, we will say that the photons were all "$T_{90°}$-polarized." The difference is a subtle one, but its significance will become apparent shortly. Now imagine that we had started with a collection of 0°-polarized photons instead. In this case, a 90°-oriented filter would have led to a "Stop" (S) result every time. Again, there is nothing new here. But how should we name the initial polarization state of the photons that produced these results? We began by referring to the photons as "0°-polarized," but the filter was set to measure polarization along 90°. Rather than identifying their initial state according to a measurement we didn't perform, let us label this state according to the measurement we *did* perform. Thus we will say that this group of photons carried the "S" value for the property of 90°-polarization, i.e., that they were all "$S_{90°}$-polarized." Wanting a way of talking about polarization that fits the particulate nature of photons leads us to think of different filter angles as measuring *different properties* of the photon. In our classical model, then, photons do not have one polarization à la Figure 3.1. Rather, they each have a definite polarization value, T or S, for every possible measurement.

A more explicit rationale for adopting this admittedly cumbersome way of describing photon polarization is in order. From a classical perspective, it makes sense to assert that whatever happens in an individual polarization measurement happens because the photon began its life with a certain (i.e., definite) property relative to the measurement actually performed. Every photon has the value T_a or S_a with regard to some possible polarization property (filter angle a). This is just a commonsense inference made by classical physicists from the nature of outcomes to the nature of properties. A definite outcome implies an initially definite property (in contemporary jargon this idea is called "faithful measurement"). Going forward we will presume along with classical physicists that every photon carries at all times one of the two possible polarization values (T or S) for *every possible* polarization measurement angle a, i.e., every possible polarization property (see Figure 3.3). The content of this presumption is effectively the principle of property definiteness (cf. Shimony 2001, 5; Zajonc 2003, 3).[7] A two-property description like "$T_{90°}, S_{45°}$" might seem to flout property definiteness, but it does not. The values $T_{90°}$ and $S_{45°}$ are not mutually exclusive, classically speaking, because they relate to *different* measurements and are thus to be associated with different properties of the photon.

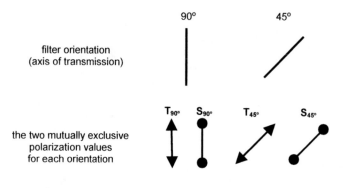

possible combinations of 90°-polarization value and 45°-polarization value for a single photon

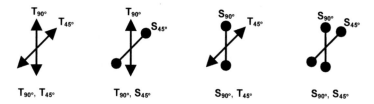

Figure 3.3   Photon polarization and property definiteness. The principle of property definiteness constrains the depiction of photon polarization, such that any given photon must always have one of two possible polarization values, T_a or S_a, for any filter orientation a. The two values are mutually exclusive and exhaustive. The top half of the figure shows two different filter orientations (90° and 45°), along with the two possible polarization values carried by a single photon for each orientation. The bottom half of the figure shows the four possible combinations of 90°-polarization value and 45°-polarization value for a single photon.

Now, it may be the case that some particular relation exists between these different polarization properties. A robust theory of photon polarization would provide us with a general account of such relations, but we do not need to develop a robust theory here. We only need to develop a minimal account of polarization that accords with the particulate nature of photons and abides by our classical intuitions. This means that for the remainder of the chapter we will assume, in accord with property definiteness, that each photon carries with it a distinct and definite polarization value for each possible polarization measurement even before any measurement occurs.

Let us now turn to the principle of "state separability." At the root of this principle is the notion of spatial separation (or what is commonly called "distance"). Physical objects that exist in separate places do so separately from one another and must therefore have separately describable states.[8] It sounds almost too obvious to mention, but it is a fundamental presumption that needs to be articulated. Einstein expressed his understanding of this principle in the following way: "An essential aspect of . . . physics is that [physical objects] lay claim, at a certain time, to an existence independent of one another, provided these objects 'are situated in different parts of space'" (Born 1971, 170). Separable states are commonly referred to as "product" states because they can be expressed as the mathematical product of two distinct state descriptions, each of which resolves to a claim about a single object. A concrete example will help clarify the idea.

Let us consider a physical system composed of two photons, A and B. In a fully state-separable world one can always say that photon A *over here* has its own α-polarization state A, say, $A = T_\alpha$, and that photon B *over there* has its own β-polarization state B, say, $B = T_\beta$. The subscript α indicates that we are talking about the polarization state of photon A, whereas the subscript β indicates that we are talking about the polarization state of photon B. The joint system composed of these two photons—we will call it C—will then have a joint $\alpha\beta$-polarization state, C, that is fully "partitionable" or "separable" in terms of the individual polarization states of A and B. Mathematically speaking, C will be "factorizable" in the sense that it can be factored into an expression where all of the information about photon A is grouped separately from all of the information about photon B. In the present example, this is the case trivially because the state of system C can be written as $C = A \otimes B = T_\alpha \otimes T_\beta$. We use a special symbol (in this case, "$\otimes$") to remind ourselves that whatever appears immediately to its left will always refer to photon A and immediately to its right to photon B. (The expression "$T_\beta \otimes T_\alpha$" is not licit because we have stipulated that β refers to the polarization state of photon B, which must always appear on the right, and that α refers to the polarization state of photon A, which must always appear on the left.) The presence of only one $\otimes$ in the description of this joint-polarization state marks it as a "product" state, i.e., a joint state whose individual states are fully separable (the individual state of each photon can be written separately). To find the likelihood of system C being in the joint state $T_\alpha \otimes T_\beta$ we simply multiply the likelihood of photon A being in T_α with the likelihood of photon B being in T_β—hence the designation "product state."

What might a nonseparable joint state look like? Here is one:

$$\left(T_{90°} \otimes S_{90°} \right) + \left(S_{90°} \otimes T_{90°} \right). \qquad 3.1$$

This state contains the sum of two distinct product states. It once again refers to two photons, A and B, but now each photon is in a more complicated combination of states $T_{90°}$ and $S_{90°}$. Most importantly, the individual state of photon A cannot be written separately from the individual state of photon B (i.e., we cannot separate their states' descriptions such that everything having to do with photon A appears to the left of a single

$\otimes$ and everything having to do with photon B appears to the right of the same $\otimes$). Classically speaking, the state given by Equation 3.1 makes no sense; $T_{90°}$ and $S_{90°}$ are classically immiscible states for a single photon. (In quantum mechanics, on the other hand, such a combination of states *is* licit—it is called a "superposition" state. In this chapter we are investigating the classical worldview, so let us not get ahead of ourselves.) Note that the state given by Equation 3.1 violates property definiteness by assigning more than one classical state to photon A and more than one classical state to photon B. Note as well that two instances of the symbol $\otimes$ are present instead of one. This final fact is what makes the state nonseparable. It cannot be reduced to a mathematical expression in which everything about photon A's polarization state resides to the left of a single $\otimes$ and everything about photon B's polarization state resides to the right of the same $\otimes$. In the language of mathematicians, this joint state is "nonfactorizable"; the states of the two photons are inextricably linked to one other. Such nonfactorizability is what led Erwin Schrödinger to use the term "entanglement" in his early reflections on the significance of quantum theory's use of superposition (1935, 555; 1983; for the view that Schrödinger took state nonseparability and cause locality to be distinct issues, see Scarani 2006, 76–77).

Here is an important caveat to the discussion in the previous paragraph. Separable (i.e., factorizable) joint states can appear quite complicated when expressed in expanded form. Consider the following joint state (again, for two photons, A and B):

$$C = \left(T_{90°} \otimes T_{90°}\right) + \left(T_{90°} \otimes S_{90°}\right) + \left(S_{90°} \otimes T_{90°}\right) + \left(S_{90°} \otimes S_{90°}\right). \qquad 3.2$$

This joint state is the sum of four distinct product states. Although it looks both indefinite and nonseparable, it can be factorized and written more simply as:

$$C = \left(T_{90°} + S_{90°}\right) \otimes \left(T_{90°} + S_{90°}\right). \qquad 3.3$$

The state is indeed indefinite, but it is not nonseparable. If the equality of Equations 3.2 and 3.3 is a puzzle, recall the basic algebraic idea of "expansion": $(a+b)\times(c+d) = (a\times c)+(a\times d)+(b\times c)+(b\times d)$. From this one can see that in the expressions above, the symbol $\otimes$ functions as a multiplier akin to "$\times$", but here we are joining (multiplying) states rather than numbers. The state given in Equations 3.2 and 3.3 is separable, i.e., everything about photon A can be made, through factorization, to appear on one side of a single $\otimes$ and likewise for photon B. This state violates the principle of property definiteness because each photon apparently carries a combination of the classical immiscible values $T_{90°}$ and $S_{90°}$—but that is a separate matter.[9]

We can now finally turn to the principle of cause locality. Thanks in no small part to the empirical success of Einstein's special theory of relativity, physicists have come to regard the speed at which all photons travel in empty space (i.e., in a true "vacuum") as being exactly the same, an unchanging or "invariant" quantity—one of a handful of nature's basic physical constants such as the charge carried by a single electron. Unlike any other speed, this speed is "absolute." Photons traveling freely in space never speed up or slow down—they are forever constrained to travel at the same, unalterable (and, truth be told, unfathomable) speed of roughly 1 billion kilometers per hour or, in the usual shorthand, "c." To understand how strange a claim this is, consider the following example. Car A is traveling down a straight road at 60 kph. From the perspective of car B, which is also traveling down the road in the same direction and at the same speed, A will appear to be standing still (there is no relative motion between the two cars). There is also a third car, C, traveling down the same road at the same speed but in the *opposite* direction. From the perspective of this car, A will appear to be moving at 120 kph. This difference reflects the basic fact that the speed at which one measures car A to be traveling will be relative to the "reference frame" within which it is measured. Surprisingly, the situation is entirely different with light. When we attempt to measure the speed of a photon coming from the headlight of car A, we will conclude that the photon is traveling at c regardless of which car we are in, B or C. The speed of light is absolute in this sense. It also imposes an "absolute limit" on the speed of any physical object or process. All massive objects are constrained to move at "subluminal" speeds (below c), whereas photons and other massless particles are confined to travel at the "luminal" speed, c.

In its pre-Einsteinian formulation, the principle of cause locality was that all physical causal interactions between separated objects are mediated by cause-and-effect relations that propagate at some finite speed contiguously through every intervening point in space. This principle depends upon the reductionism inherent in state separability, for there would be no reason to think that all physical interactions must be mediated solely by "local" cause-and-effect relations without an understanding of nature as composed of distinct and thus conceptually separable parts ("this" one and "that" one) having their own intrinsic properties. The roots of the principle of cause locality lie in the atomism of the Greeks, the lasting influence of which can be seen in early modern science's understanding of physical change in terms of the notion of "contact action" (Howard 1989, 243). In his *Physics*, Aristotle divided motion into two kinds: self-motion and contact motion. The former is characteristic of living things, whereas the latter is characteristic of inanimate objects. Aristotle thought of action in terms of the position in space of the agent doing the moving: "Agents simply cannot push or pull where they themselves are not bodily present" (quoted in McMullin 1989, 276).[10] The principle of cause locality does not mean that distant events cannot influence each other, only that when they do so their influence must be conveyed or mediated by cause and effect through every point in between at some finite speed. Descartes inherited and embraced Aristotle's notion of contact action, making it a central principle of his mechanistic philosophy. What had been Aristotle's heavenly spheres became Descartes' heavenly ether composed of countless whirling vortices. The opposite of contact action is typically referred to as "action-at-a-distance," meaning physical change that is not mediated through direct contact. This was anathema to classical physicists. The fact that Newton's theory of gravity seemed to require action at a distance was a source of great consternation, both to Newton and to his opponents. Recall that when Einstein derided entanglement, he referred to it as "spooky action at a distance" (Born 1971, 158).

Einstein's special theory of relativity brought new specificity to the principle of cause locality. Whereas the pre-Einsteinian account simply limited the propagation of all causal processes to some finite speed (*pace* Newton), Einstein identified this speed as the speed of light, c. This implies that if a photon traveling 1 billion kilometers per hour cannot span the "distance" in space and time between two spatiotemporal events, then nothing can. The two events are said to be "space-like" separated from one another. From the relativistic point of view, space-like separated events are causally isolated from one another—neither can have any effect on the other.[11] Here is how Einstein expressed the principle of cause locality (which he called "contiguity"):

> The following idea characterizes the relative independence of objects far apart in space (A and B): [an] external influence on A has no direct influence on B; this is known as the "principle of contiguity," which is used consistently in field theory. (Born 1971, 171)

The apparently instantaneous character of many modern forms of communication is in fact a felicitous illusion stemming from the extreme speed of electromagnetic signals. No communication can really be instantaneous, strictly speaking, in a world where the principle of cause locality holds.

3   OBJECTING TO QUANTUM THEORY

The founders of quantum theory, especially Albert Einstein and Niels Bohr, debated its adequacy, not on empirical grounds but on conceptual grounds: Did the theory tell the whole story about the microphysical world? Einstein refused to accept that quantum theory was a conceptually complete account of the processes it described. He saw as clearly as anyone what the theory implied about physical processes. Although his famous quip, "God does not play dice," was directed at the apparent indeterminism of the theory, a number of his other critical comments—including the paper he coauthored with Boris Podolsky and Nathan Rosen (EPR 1935)—were aimed more directly at the apparent entanglement of objects implied by the theory.[12] Upon reading the EPR paper, Schrödinger called entanglement "*the* characteristic trait" of quantum theory (1935, 555; italics in original; cf. Scarani 2006, 76–77).

Einstein recognized the tension between quantum theory and the classical worldview even when the theory was still in its infancy. He appears first to have voiced his concerns in 1927 at a conference of the Solvay Institute in Brussels, just a year after Schrödinger and Heisenberg had constructed their distinct mathematical approaches to the theory. This was the fifth in a series of conferences attended by many of the physicists who were involved in the development of the new quantum theory. Einstein argued in Brussels that if a superposition were to be regarded as a complete description of an individual particle rather than as a statistical description of an "ensemble" of particles,[13] then a particle that was initially prepared to be in a superposition of different locations could be said to transition instantaneously upon measurement from some kind of global presence to a localized presence at a particular point. This transition appeared to him to contradict cause locality because it happened with reference to all points in space instantaneously and without any mediating causal connection. Einstein's colleague Paul Ehrenfest remarked in 1932, "If we recall what an uncanny theory of action-at-a-distance [*unheimliche Fernwirkungstheorie*] is represented by [superposition], we shall preserve a healthy nostalgia for a four-dimensional theory of contact-action [*Nahwirkungstheorie*]!" (quoted in Jammer 1974, 117).

At the Seventh Solvay Conference in 1933 Einstein posed a thought experiment to Léon Rosenfeld, one of Bohr's assistants. Begin with two particles that have identifiable momenta. The particles interact (i.e., come into contact) briefly and then separate. If one measures the exact momentum of one particle, one can calculate the exact momentum of the other. If, on the other hand, one had chosen to measure the exact position of one particle, then one could calculate the exact position of the other. But if, according to the standard interpretation, neither particle can have both an exact (i.e., definite) position and an exact (i.e., definite) momentum, it would seem that one's measurement of one particle influences the other. Einstein, according to Rosenfield (1983, 137), then asked, "is it not very paradoxical? How can the final state of the second particle be influenced by a measurement performed on the first, after all physical interaction has ceased between them?" To understand why Einstein found this idea "paradoxical"—granting that it is impossible to know whether Einstein actually used the word in his conversation with Rosenfeld—one needs to understand something more about what quantum theory says. Position and momentum, according to the theory, are "noncommuting" properties, which means that they cannot both be measured at the same time. According to the standard interpretation, this means that the position and momentum of a quantum particle cannot be simultaneously absolutely definite.[14] But as Einstein pointed out, by choosing to measure the position or the momentum of one particle one apparently causes the same property of the distant particle to become definite—without ever interacting directly with it! One sees here the germ of the EPR paper.

As an unflagging proponent of cause locality, Einstein objected to what he saw as a peculiar holism within quantum theory. His concern appears to have centered on the role of superposition within the theory. He recognized that if one applied the notion of superposition to an individual particle, thereby allowing its properties to have fuzzy, indefinite values, the particle's transition to having a definite property upon measurement happened instantaneously. Einstein regarded this as highly problematic and argued instead that the concept of superposition should be understood as a statistical description of the behavior of ensembles, not a complete description of individual particles. If quantum theory seemed to entail indefiniteness, this was only because the mathematics was being interpreted incorrectly on this point. Particles must have definite properties at all times, he thought, even when they are not being observed. If one could get behind the "fuzziness" of quantum theory to the definiteness of reality, then the strange influences implied by the theory would probably disappear as well—or so Einstein thought.

In 1935 EPR published their now-famous critique of quantum theory (EPR 1935).[15] In line with the argument Einstein had previously presented to Rosenfeld in 1933, they pointed out that within quantum theory separated systems can continue to influence one another and can do so no matter how widely separated they might become. They found this idea so implausible that they made it the centerpiece of their argument. Instantaneous influences between distant objects cannot be correct, so any idea that implies such influences—e.g., the indefiniteness of properties—can't possibly be correct either. EPR expressed confidence at the end of the paper that some additional consideration not yet present within the quantum account would reinstate property definiteness for all of the actual properties of physical objects. Although the argument did not initially motivate the physics community to hunt for a more complete theory—David Bohm did later return to the issue (1952)—EPR deserve lasting credit for pointing out early on that quantum theory implies something very odd and even troubling, at least from a classical point of view, about the relationship between spatially separated objects.

It is not entirely clear from Bohr's writings, which are notoriously cryptic, whether he ever fully understood Schrödinger's point about entanglement setting quantum theory apart from all of classical physics (but see Howard 1990; Barad 2007). In any case, Bohr appears to have recognized the holistic quality of the measurement process inasmuch as he responded to EPR by calling their attention to the influence of the global experimental setup "on the very conditions which define the possible types of predictions regarding the future behavior of the system" (1983, 148). Some years later Bohr (1987, 2) commented that Planck's discovery of the elementary quantum of action had revealed a wholeness within quantum processes. Although it is difficult to say with any confidence precisely what he thought about entanglement, throughout his career he continued to resist what he perceived to be Einstein's too-classical instincts.

EPR showed that the idea of property indefiniteness led to entangled behavior. A measurement in one part of space could have an instantaneous effect in distant parts of space, which seemed to imply a violation of cause locality. But did this observation about the character of quantum theory imply anything peculiar about the actual observable behavior of quantum particles? Would there ever be any empirical evidence to support the reality of these instantaneous effects? In 1948 Einstein commented that nothing about the actual behavior of the physical world required him to abandon the idea that physical objects in different parts of space exist independently of one another (Born 1971, 172–173). In 1964, nine years after Einstein's death, the possibility of such behavior appeared in the groundbreaking work of John Bell.

4   FROM OBJECTION TO PREDICTION

In a brief article appearing in the first volume of the short-lived journal *Physics*, Bell (1983) called attention to a previously unnoticed result—a mathematical inequality—buried within the details of the EPR argument, as later simplified by David Bohm (1989, chap. 22; I refer to Bohm's version as "EPR–Bohm" and to Bell's version simply as "Bell."). The point of Bell's article was to develop a prediction based on classical presuppositions and then to point out that this prediction cannot be reconciled with the prediction generated by quantum theory. The focus in this chapter, as I have already stressed, is on the classical element within the argument, which is to say, on the theorem Bell derived using Einstein's articulation of a "classical" viewpoint.

Bell's paper presented an elegant example of what physicists call a "thought experiment." The act of thinking through an experiment without actually performing it can sometimes be very helpful when it comes to highlighting implications of a line of thought, shedding light on a problem when an actual experiment isn't feasible, or checking the consistency or ramifications of a novel idea. All three virtues apply to Bell's modification of EPR–Bohm's thought experiment. What Bell discovered through his particular thought experiment was a novel implication of the classical worldview. This implication took the form of a mathematical inequality that was testable, at least in principle. Quantum theory, on the other hand, generated a prediction that did not obey the inequality Bell had derived. One often hears today that quantum theory "violates" Bell's inequality. By this it is meant that the Einsteinian "classical" view leads to one prediction and quantum principles to another. Figure 3.4 shows the three-way

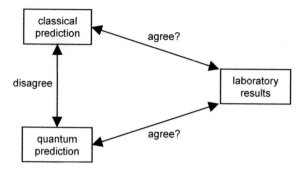

Figure 3.4 Relationships among classical prediction, quantum prediction, and laboratory results. The classical prediction and quantum prediction disagree about what should happen. Do laboratory tests confirm one prediction or the other, or neither of them? The current chapter presents the classical prediction.

relationship between the classical prediction, the quantum prediction, and laboratory results.

Whereas Einstein had earlier hoped to rescue quantum theory from entanglement by ridding it of indefiniteness, Bell wanted to ascertain whether one could deploy EPR's strategy and still hope to produce a theory that was capable of matching the predictions of quantum theory. EPR had shown that indefiniteness led to entanglement, and so the only hope for an un-entangled theory appeared to lie in a return to property definiteness. Would a theory that embraced such definiteness be able to reproduce all the results of quantum theory? David Bohm had already shown that this was possible (1952), but at a surprising cost. His own interpretation manifested something like entanglement with a vengeance—Bell called Bohm's approach "grossly nonlocal" (1983, 403). In light of Bohm's work, Bell wondered whether it might be possible to construct a theory that would match the quantum prediction and could still be interpreted as supporting property definiteness without the ontological "excesses" of Bohm's approach. What Bell discovered, in the form of his famous inequality, was that such a theory is *not* possible. No property-definite theory can match the quantum prediction, except if it contains something like entanglement—as Bohm's theory had.

The genius of Bell's argument lay in two important insights. First Bell translated EPR's classically formed intuitions about things like property-definiteness into the language of mathematics. Second, he made a crucial change to the EPR–Bohm setup that allowed him to expose the predictive conflict between quantum theory and any nonentangled, property-definite theory (in the literature such a theory is commonly referred to as a "local hidden-variables theory"). The change Bell made was to consider the results that would obtain if one were to measure one property of one object but a *different* property of the other object. In the introduction to his paper Bell claimed that the question of cause locality was the essential difficulty in the

Bohr–Einstein dispute.[16] He also argued (1983, 403), contrary to popular opinion, that the causally nonlocal character of Bohm's hidden-variables version of quantum theory was not the liability it appeared to be because causal nonlocality must appear in any theory matching the predictions of quantum theory. To better appreciate Bell's work we now turn to the task of familiarizing ourselves with a Bell-type thought experiment.

5   A THOUGHT EXPERIMENT

The apparatus that is required to run a basic Bell-type experiment has three main components: two measurement devices and a photon source positioned between them (see Figure 3.5).[17] During each run of the experiment the source produces a single pair of photons A and B with a particular set of polarizations.[18] The two photons fly off in opposite directions toward their respective measurement devices, each of which is composed of a polarization filter with a rotatable axis of transmission oriented in some direction (angles α and β, respectively), perpendicular to the photons' common line of flight, and a photon detector. The positive horizontal axis viewed from the right side of the apparatus is arbitrarily designated as $0°$, all angles being measured counterclockwise from this axis. The measurement process in each wing begins with the filter transmitting (T) or stopping (S) a photon and concludes with the detector noting the filter's orientation and recording A_T (or B_T) if the photon reaches the detector and A_S (or B_S) if it does not.

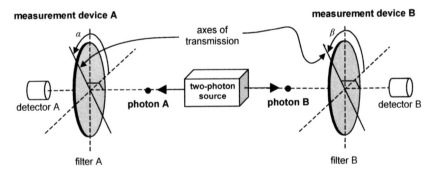

Figure 3.5   Bell apparatus with two-photon source and two measurement devices. Each measurement device is composed of a polarization filter and a photon detector. The source produces pairs of photons, one pair for each run of the experiment, imparting to each pair a set of polarizations (not shown) lying in the plane of the filters. The transmission axis of each filter can be set to any angle. The photons' polarizations lead either to correlated or to anti-correlated joint outcomes—transmit (T) or stop (S) for each photon—as each photon is detected, or not, by its own detector. The joint outcomes of many runs can then be used to calculate the source's average preference for correlations versus anti-correlations under various combinations of filter angles. A particular choice of filter angles constitutes a specific "joint measurement scenario."

Our particular interest in this apparatus lies not in the individual polarization measurements made by device A or B, but in the measurements of A and B together, i.e., in the so-called "joint outcome" of a single run of the experiment using a particular set of filter orientations. Any given set of filter orientations will be referred to in the following as a "joint measurement scenario." The basic building blocks of Bell's theorem are the four possible joint outcomes: $A_S B_S$, $A_S B_T$, $A_T B_S$, and $A_T B_T$. To avoid confusion in what follows, I will always use the terms "individual" and "joint" when referring to the *type* of outcome within a single run and the terms "single" and "many" when referring to the *number* of runs (typical laboratory tests of Bell-type inequalities involve many thousands of runs). One can associate each joint outcome with a numerical value by arbitrarily assigning the following values to individual outcomes: $A_S = B_S = -1$ and $A_T = B_T = +1$ (stop is assigned -1 and transmit is assigned $+1$). By multiplying all possible combinations of A and B outcomes together, one arrives at the following numerical values for joint outcomes: $A_S B_S = A_T B_T = +1$ and $A_S B_T = A_T B_S = -1$. In accord with convention, I refer to these two types of joint outcome as "correlation" $(+1)$ and "anti-correlation" (-1). Whether a particular instance of correlation stems from $A_S B_S$ or $A_T B_T$ (or a particular instance of anti-correlation stems from $A_S B_T$ or $A_T B_S$) is not relevant to Bell-type arguments and will not concern us here. What will concern us is a mathematical description of the tendency of the source to produce one type of joint outcome or the other, correlation or anti-correlation, under a particular joint measurement scenario.

Because of the apparent randomness of individual outcomes, one can never predict whether a single pair of photons will correlate or anti-correlate on any given run. A single joint outcome reveals what kind of polarizations the source did impart to the pair of photons with regard to a particular joint measurement scenario, but it tells us nothing about the source's *tendency* to prefer one type of joint outcome or the other for this scenario. To get a clearer picture, we need to measure the polarizations of a large ensemble of photon pairs under the joint measurement scenario and then examine the overall number of correlations and anti-correlations that occur in the ensemble. By adding joint outcomes together and dividing the overall result by the number of runs N, we can describe the source's tendency to prefer correlations or anti-correlations for this particular joint measurement scenario in terms of the average of joint outcomes. I will refer to this as the source's "average preference for correlation versus anti-correlation" or, more compactly, its "average preference."[19] If the source always produces anti-correlated pairs for a particular joint measurement scenario, its average preference for that scenario will be -1. On the other hand, if it always produces correlated pairs, its average preference for that scenario will be $+1$. If the source produces anti-correlated pairs some of the time and correlated pairs otherwise, the average preference for the scenario will lie somewhere between -1 and $+1$. Finally, if anti-correlations and correlations are produced by the source in equal numbers, the average preference for

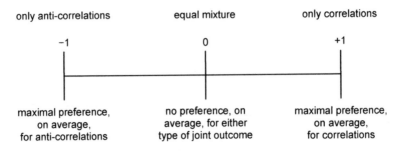

Figure 3.6 Average preference under a single measurement scenario. The range of possible values for the average of joint measurements performed on an ensemble of photon pairs under some particular joint measurement scenario is -1 to +1. The particular value obtained for any given ensemble indicates the "average preference" displayed by the pairs in the ensemble, i.e., the degree to which the source favors one type of joint outcome over the other under this particular joint measurement scenario. An average preference of 0 indicates equal numbers of correlations and anti-correlations.

the scenario will be zero. Figure 3.6 shows the range of possibilities. As we shall see in the following, this particular method of tabulating correlation leads to a limit on the amount of average preference that can be displayed by an ensemble of photon pairs under multiple joint measurement scenarios, assuming property definiteness, state separability, and cause locality.

How did Bell use joint outcomes to derive his inequality? Bell's crucial insight had to do with the orientations of the two polarization filters. Whereas EPR and Bohm had considered only joint measurement scenarios in which the filters were oriented along a common angle, Bell asked himself what would happen when they were oriented along *different* angles. He allowed each filter to assume one of three possible orientations (α, β, and γ) during each run. This led him to an inequality that took the form of $x \leq 1$, where x is shorthand for a larger mathematical expression that combines the average preference values for three specially chosen joint measurement scenarios among the nine possible combinations of detector settings: α for filter A and β for filter B, α for filter A and γ for filter B, and β for filter A and γ for filter B.[20] Bell showed that the quantum prediction violates his inequality, i.e., differs from the classical prediction, when the differences between α, β, and γ become large.

In the version of the thought experiment presented here, I will give each polarization filter its own set of two possible orientations: α_1 and α_2 for filter A, and β_1 and β_2 for filter B. The reason for this slightly more complicated arrangement is that it is experimentally very difficult to implement Bell's scenario, because one has to make sure that the angles on the left side of the apparatus are exactly the same as those on the right side. (Practically speaking, setting two filters to *exactly the same* angle is an impossible task.) In the present case, no such constraint is supposed. One can freely choose any two angles on the left and any two on the right. This leads to four possible joint measurement scenarios for any given run: $\alpha_1\beta_1$, $\alpha_1\beta_2$, $\alpha_2\beta_1$, and $\alpha_2\beta_2$. To be clear, the angles represented by α_1, α_2, β_1, and β_2 can all be different from one another—one can assign whatever angle one wants to each letter.

It is important here not to confuse the concept of "joint measurement scenario" with the concept of "joint measurement outcome" introduced earlier. Any of the four possible joint outcomes can appear under a given joint measurement scenario; outcomes are *results*, whereas joint measurement scenarios are the *experimental arrangements* under which those results obtain. Unfortunately, the fact that there are four of each opens the door to confusion. Later, we will compare outcomes across different joint measurement scenarios, but for now let us focus on scenario $\alpha_1\beta_1$. To anticipate the final result, the classical prediction derived in the following takes the form $x \leq 2$ where x is once again shorthand for a more complex expression that combines the average preference values for all four joint measurement scenarios.

When physicists measure the polarization of jointly produced photons in the laboratory according to the procedure described in the following, they find two important results. First, when they consider individual outcomes, no discernable pattern appears on either side of the apparatus, regardless of which joint measurement scenario is chosen. Second, pairs of photons display a clear tendency to correlate or anti-correlate, depending on the chosen measurement scenario—a surprising result in light of the first. Specifically, they tend to anti-correlate more as the difference between α and β approaches $0°$ and to correlate more as this difference approaches $90°$. Generally speaking, however, joint measurements lead randomly to both types of joint outcome and therefore to less-than-maximal average preference, with the average preference falling to zero as the difference between α and β approaches $45°$. In sum, jointly produced photon pairs display an uncanny mix of apparently random and nonrandom behaviors. Jointly (and collectively) their behavior points to a source with varying preferences for correlation and anti-correlation, even though individually this same behavior appears completely random. I will say more in Chapter 5 about this important feature of entanglement.

How might we account for this curious combination of behaviors from a classical point of view? Perhaps the appearance of various amounts of correlation and anti-correlation is just a "matter of statistics." In other words, perhaps it is only a consequence of the limited mix of initial polarizations the source has at its disposal. Prior to Bell, most physicists probably would have invoked the statistical account of quantum behavior given by quantum theory and left it at that. But Bell demonstrated that only a certain amount of average preference is compatible with a classical point of view and, furthermore, that the quantum prediction can *exceed* and thus *violate* the classical limit in some cases. The discussion so far has focused on outcomes, but it is important to remember that everything said in the preceding presumes that the photons' actual, definite polarization states are the primary reason for the outcomes that obtain and are in no way a product of the actual measurement process influencing these states.

In the present Bell experiment, we must be sure that the measurement devices will accurately perform their measurements without actively influencing the results. Although we will simply assume that the filters and detectors all function properly *qua* filters and detectors, there is a further and perhaps odd-sounding concern to consider, namely, that the filters might somehow communicate with the photon source or with each other at the beginning of each run by some local (i.e., subluminal or luminal) mechanism, which could alter the amount of correlation measured. In fact this problem was recognized by Yakir Aharonov and Bohm (1959) and reiterated by Bell (1983). Three separate questions can be raised. First, can a measurement device send local signals *to the source* regarding its configuration ahead of the actual run, thereby possibly exerting some measure of influence over the source's choice about which polarization states to impart to the photons? Second, can one measurement device send a signal *to the other measurement device* ahead of time, thereby possibly exerting some measure of influence over the distant outcome? Finally, can a measurement device send a signal *to the photon in the other wing* before that photon's polarization is measured? Granted, the thought of a measurement device broadcasting a signal about its configuration, or of the source, the distant photon, or the distant measurement device responding in any sort of coordinated way seems highly unlikely—not to mention overly anthropomorphic—but there is a real issue to worry about here. The point is not to impute intentionality to any part of the apparatus but to be absolutely sure that the measurement devices cannot communicate—by whatever obscure or unanticipated but nonetheless physical and luminal or subluminal means—their own configurations to the source, the distant photon, or each other in a timely manner. We don't need to know how this might happen in general. We only need to make sure that it *cannot* happen in our thought experiment. Otherwise, the door would be open for critics to allege the possibility of a "local" explanation of an experimental violation of our Bell-type inequality. This problem is commonly referred to as the "locality loophole."

Consider what would happen if either measurement device were able to communicate its configuration to the source before the latter had produced its photons for a given run. The source would then be in a position to adjust the photons' polarization states in anticipation of the measurements to be made. Note that timely communication is not hard for a measurement device to achieve. The signal about its filter setting only needs to leave the measurement device *after* its configuration has been set and to arrive at the source *before* it has produced its pair of photons for that run. Fortunately, there is a simple way (conceptually, at least) to prevent this from happening:

Delay setting each filter's orientation until neither a subluminal nor luminal signal could reach the source (or the other measurement device) in a timely fashion. Because we are assuming by the principle of cause locality that superluminal signaling is impossible, this procedural adjustment—call it "orientation delay"—guarantees that no signal can be sent by either measurement device to the source before it has produced its photons or to the distant measurement device before it has measured its own photon. The third possibility is also ruled out because no signal will be able to catch up to the distant photon after it leaves the source (the distant photon is traveling at c, of course, which means that nothing can overtake it, not even another photon). Any correlation present between the polarization measurements must then be due solely to the photon source acting by itself.

Before turning to the actual derivation of Bell's inequality, let's review what we have covered in the chapter thus far. We began by examining the concept of the photon, including its particulate nature, its polarization, and its speed. These concepts allowed us to explain in detail how classical physicists understood property definiteness, state separability, and cause locality—the three properties by which any classical account of photon polarization must abide. We considered some of the objections registered against quantum theory, particularly those of Einstein, and then familiarized ourselves with the basic outline of Bell's thought experiment. This section introduced the basic arrangement of the photon-measurement apparatus and established how one assigns numerical values to the various individual and joint-polarization outcomes that can occur within a single run under any given joint measurement scenario. The notions of "correlation" and "anti-correlation" enabled us to quantify the tendency of the photon source to prefer, on average, one type of joint outcome or the other for a given joint measurement scenario; and the concept of an "average outcome of joint-polarization measurements" provided a straightforward measure of this preference. Finally, the idea of orientation delay provided the necessary guarantee that a measurement device could not exert a timely influence over the source, the distant measurement device, or the distant photon.

6   A CLASSICAL INEQUALITY

The derivation presented in this final section occurs in two stages, each of which begins with a simple algebraic statement. The first statement leads to an intriguing equality, whereas the second transforms the equality into a classical, Bell-type inequality.[21] Property definiteness and state separability play important roles in the first stage, whereas cause locality plays an important role in both stages. We begin with the following general mathematical expression:

$$Q = x_1 y_1 + x_1 y_2 + x_2 y_1 - x_2 y_2 \qquad\qquad 3.4$$

or equivalently,

$$Q = x_1(y_1 + y_2) + x_2(y_1 - y_2)$$ 3.5

What is surprising about this relatively simply algebraic expression is that if one confines each of the four variables on the right-hand side to ±1, then the value of the entire expression, Q, is constrained to ±2:[22]

$$Q = x_1(y_1 + y_2) + x_2(y_1 - y_2) = \pm 2$$ 3.6

Now, recall that we assigned the various individual outcomes of our Bell-type experiment the values ±1. Can we construct an expression analogous to Equation 3.4 with the relevant versions of our placeholders for measurement outcomes $A_k(\alpha_l \beta_r)$ and $B_k(\alpha_l \beta_r)$? Here the subscript k indicates a particular run of the experiment, i.e., a measurement of the kth pair of photons A_k and B_k. Let us substitute in appropriately for l and r to produce the four joint measurement scenarios ($l = 1$ and $r = 1$, $l = 1$ and $r = 2$, $l = 2$ and $r = 1$, and $l = 2$ and $r = 2$), which gives us the following physically meaningful expression:

$$A_k(\alpha_1 \beta_1) B_k(\alpha_1 \beta_1) + A_k(\alpha_1 \beta_2) B_k(\alpha_1 \beta_2) + A_k(\alpha_2 \beta_1) B_k(\alpha_2 \beta_1)$$
$$- A_k(\alpha_2 \beta_2) B_k(\alpha_2 \beta_2).$$ 3.7

This expression includes eight individual outcome placeholders and four joint outcome placeholders. For example, one of the individual placeholders is $A_k(\alpha_1 \beta_1)$ and one of the joint placeholders is $A_k(\alpha_1 \beta_1) B_k(\alpha_1 \beta_1)$. Each individual placeholder stands for +1 or –1, as does each joint placeholder (because the values of paired individual placeholders are multiplied to obtain the value of the corresponding joint placeholder). The subscripted numbers on the Greek letters replacing l and r allow us to identify the particular joint measurement scenario to which each individual polarization value is related. All four joint placeholders refer to the *same* pair of photons (the kth pair, A_k and B_k), so we will call this expression a "single-pair sum."[23]

How should we interpret the physical meaning of Equation 3.7? We could regard each joint term as a joint outcome, but then we would be forced to contend with the awkward fact that the kth pair of photons can only ever be measured under one joint measurement scenario. Adding together the values for four equally possible but mutually exclusive joint outcomes seems like a dubious place to begin the derivation.[24] The problem can be avoided by interpreting these terms as placeholders for the definite joint-polarization *states* of the kth pair of photons, regardless of whether these states are ever measured. The overall expression can then be interpreted as a sum of the numerical values associated with these states carried by a single pair of photons. Property definiteness warrants the conclusion that the kth pair of photons must be in some definite polarization with regard to *every possible* joint-polarization measurement, i.e., that each photon within the pair must carry along some definite polarization value for *every possible* filter orientation so as to be prepared to yield a meaningful measurement for whatever orientation it encounters.[25] State separability is crucial as well because *non*separable states (e.g., Equation 3.1) lead to property *in*definiteness in the context of cause locality (about which more in a moment). Although the unavoidable destruction of each pair of photons during a single joint measurement prevents this kind of sum from ever being tested, one can nonetheless ascribe a definite value to each individual property placeholder and thus to each joint property placeholder as well.[26] Recalling that each of the placeholders has a value of ±1, one concludes that this sum must have some definite numerical value.

At this point it is tempting to equate Expression 3.7 with Equation 3.6, but this will not work because the terms in Expression 3.7 cannot be factored. This is because each term depends not only on the setting of the near filter but on the setting of the distant filter as well. We cannot presume that $A_k(\alpha_1\beta_1)$ and $A_k(\alpha_1\beta_2)$ have the same value because the two terms are linked to the distant filter under *different* joint measurement scenarios. Here the principle of cause locality comes into play by allowing us to rule out any possible dependence of each photon's polarization state on the distant filter. With the help of cause locality we can write a simpler, "local" version of the single-pair sum in which each individual polarization value is now linked only to the setting of the near filter[27]:

$$A_k(\alpha_1)B_k(\beta_1) + A_k(\alpha_1)B_k(\beta_2) + A_k(\alpha_2)B_k(\beta_1) - A_k(\alpha_2)B_k(\beta_2) \qquad 3.8$$

The local version *can* be factored as

$$A_k(\alpha_1)\left[B_k(\beta_1) + B_k(\beta_2)\right] + A_k(\alpha_2)\left[B_k(\beta_1) - B_k(\beta_2)\right] \qquad 3.9$$

This expression has the same form as Equation 3.6 and therefore must have the same value, and so we write:

$$Q_k = A_k(\alpha_1)\left[B_k(\beta_1) + B_k(\beta_2)\right] + A_k(\alpha_2)\left[B_k(\beta_1) - B_k(\beta_2)\right] = \pm 2 \qquad 3.10$$

It is important to note that the nonfactorizability of Expression 3.7 is due to the "nonlocal" dependence of each individual placeholder on the distant filter setting. This is unrelated to the nonfactorizability of entangled states discussed earlier in this chapter. Before proceeding any further, we should stop to review the physical portion of the derivation up to this point. Expression 3.7 gave a sum of joint-polarization values under the four different joint measurement scenarios. We reinterpreted it to be the sum of values associated with the joint-polarization *states* of a single pair of photons, using property definiteness to argue that each individual property placeholder within a single-pair sum must have a definite value. Because state inseparability rules out property definiteness in the context of cause locality, we were led to invoke state separability (we will consider David Bohm's nonlocal retrieval of property definiteness in Chapter 5). Finally, thanks to the principle of cause locality, we were able to rule out the possibility of either filter influencing the production or distant measurement of polarization states by any local process. This issue must ultimately be addressed operationally in the laboratory by waiting to set the filters' orientations until no subluminal or luminal signal containing information about these orientations can reach the source or distant measurement device in a timely manner.

Now observe what happens when we take the absolute value of the unfactorized version of Equation 3.10:

$$|Q_k| = |A_k(\alpha_1)B_k(\beta_1) + A_k(\alpha_1)B_k(\beta_2) + A_k(\alpha_2)B_k(\beta_1)$$
$$- A_k(\alpha_2)B_k(\beta_2)| = 2. \qquad 3.11$$

Although the effect is not immediately obvious, this slight adjustment opens the door to an equality involving an average. Here is why. If $|Q_k|$ is equal to 2 for some k, then $|Q_k|$ must be equal to 2 for all k and therefore the average of $|Q_k|$ must also be equal to 2. We can conclude the first stage of the derivation by explicitly writing the average of $|Q_k|$, i.e., by summing Equation 3.11 over k (the number of a particular photon pair) from 1 to N (the overall number of photon pairs) and then dividing by N:

$$\frac{\sum_{k=1}^{N}|Q_k|}{N} = \frac{1}{N}\sum_{k=1}^{N}|Q_k| = 2. \qquad 3.12$$

Here I have employed the summation symbol Σ, which avoids having to write down a separate $|Q_k|$ for every k (there could be tens of thousands). We are moving ever closer to a testable expression, but we are not there yet. The value of $|Q_k|$ still cannot be experimentally determined because it is a single-pair sum.

The second stage of the derivation begins as the first did, with a simple algebraic truth having nothing to do with physics: The absolute value of the average of a set of numbers is always less than or equal to the average of the numbers' absolute values. Once again, a simple example will help clarify the idea. Given some set of numbers, say (1, –1, –1), it follows that

$$\text{Ave}(|1|,|-1|,|-1|) = 1, \qquad 3.13$$

$$|\text{Ave}(1,-1,-1)| = 1/3. \qquad 3.14$$

The first of these expressions gives the average of the numbers' absolute values, whereas the second gives the absolute value of their average. The discrepancy between the two arises because the positive and negative numbers in Equation 3.13 all contribute positively to the sum and hence to increasing the overall average, whereas the negative numbers in Equation 3.14 contribute negatively to the sum, leading to a relatively smaller average. Had the original set included only positive or only negative numbers, the values of the two expressions would have been equal. In general, the relationship named at the beginning of the paragraph holds true: The absolute value of the average of a set of numbers is less than or equal to the average of the numbers' absolute values.

Now consider Equation 3.12 again. This expression is the average of a set of absolute values. If we were to write it instead as the absolute value of an average, the resulting expression would necessarily be less than or equal to the value of the original:

$$\frac{1}{N}\sum_{k=1}^{N}|Q_k| = 2 \quad \rightarrow \quad \left|\frac{1}{N}\sum_{k=1}^{N}Q_k\right| \le 2. \qquad 3.15$$

The new expression on the right side *is* a testable inequality, although not obviously so in its present form. In fact, this expression is the Bell-type inequality we have been aiming for. A few final adjustments will render it more transparent.

First, we re-expand Q_k in terms of its four joint-polarization states:

$$\left|\frac{1}{N}\left\{\sum_{k=1}^{N}\left[A_k(\alpha_1)B_k(\beta_1) + A_k(\alpha_1)B_k(\beta_2) + A_k(\alpha_2)B_k(\beta_1)\right.\right.\right.$$
$$\left.\left.\left. - A_k(\alpha_2)B_k(\beta_2)\right]\right\}\right| \le 2. \qquad 3.16$$

Next, we separate the four sums and distribute "$1/N$" to each, transforming each sum into an average:

$$\left|\frac{1}{N}\sum_{k=1}^{N}A_k(\alpha_1)B_k(\beta_1) + \frac{1}{N}\sum_{k=1}^{N}A_k(\alpha_1)B_k(\beta_2) + \frac{1}{N}\sum_{k=1}^{N}A_k(\alpha_2)B_k(\beta_1)\right.$$
$$\left. - \frac{1}{N}\sum_{k=1}^{N}A_k(\alpha_2)B_k(\beta_2)\right| \le 2. \qquad 3.17$$

Finally, we can employ a common bit of physics shorthand for an average, "$\langle \ \rangle$," to write the expression more compactly:

$$\left|\langle A(\alpha_1)B(\beta_1)\rangle + \langle A(\alpha_1)B(\beta_2)\rangle + \langle A(\alpha_2)B(\beta_1)\rangle - \langle A(\alpha_2)B(\beta_2)\rangle\right| \le 2. \quad 3.18$$

Here at last is our Bell-type inequality for polarization measurements performed on an ensemble of photon pairs under four distinct joint measurement scenarios. This inequality is testable—assuming fair sampling (see note 26)—because each of the averages can be calculated from different subpopulations of the ensemble. Assuming further that properties always lead straightforwardly to outcomes (faithful measurement), this inequality will hold not only for properties but for outcomes as well. Data to test the inequality are gathered by measuring the polarizations of a random quarter of the overall population of photon pairs under each of the four joint measurement scenarios.

How to test this inequality in the laboratory? First, we would need to choose two possible orientations for filter A (α_1 and α_2) and two for filter B (β_1 and β_2). Then, during each run—but only *after* the source had released its photons—we would set each filter at random to one of its two possible orientations in anticipation of the photons' arrival. After gathering a large amount of data under all four joint measurement scenarios (say, many thousands of data points under each scenario), we would sort the data into four different "buckets" corresponding to the four different joint measurement scenarios. We would then calculate the first joint-outcome average using data from the first bucket, and so on, until we had obtained all four averages. Finally, we would insert these four values into Equation 3.18 to determine whether or not the inequality had been observed or violated.

In fact, this is what experimental physicists working on quantum phenomena have been doing in the laboratory since the 1970s. The reception of Bell's work and the history of experimental tests of his inequality take us to the next part of the story, in which the focus shifts from the classical viewpoint to the quantum viewpoint. The experiments that have been performed clearly *contradict* the classical prediction, but what light do they shed on the quantum prediction?

In order to tackle this question we need to answer another set of questions about quantum theory itself. How does the theory differ from the classical perspective in its characterization of physical processes? In particular,

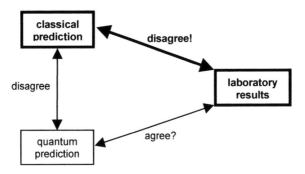

Figure 3.7 Laboratory results disagree with classical prediction. Do they confirm the quantum prediction? The following chapter examines the quantum point of view.

how does quantum theory treat the polarization of photons? And what is the quantum method for generating a prediction in a Bell-type experiment? Settling these important theoretical issues is the task of the next chapter.

The inequality rederived in the preceding reflects what many physicists and philosophers of physics take to be one of the greatest achievements of nineteenth-century physics—this, despite the fact that as a prediction it is empirically, verifiably wrong. After all, Bell's inequality has been unmistakably violated in Bell-type experiments. What then makes it such a valuable tool? Bell's inequality led to the discovery of something radically new about the world, as well as to the realization that something must be profoundly wrong with our old habits of thought. As physicists know all too well, testable differences between theories based on classical, "commonsense" ways of thinking and theories like quantum mechanics that seem to belie our classical intuitions are precious and few. But Bell managed to find one. His work sparked an experimental and conceptual revolution that overturned what physicists thought they already knew about the world, even as it put one more feather in the cap of a theory that resists interpretive consensus to this day. There is strange but beautiful irony in what Bell accomplished. Seldom in the history of physics has the careful deployment of an apparently faulty set of presuppositions been such a boon to the intellect, or to the imagination.

4   Entanglement in Quantum Physics

In 1964 John Bell demonstrated that quantum theory does not always agree with our classical intuitions about what should happen in the laboratory. Bell's uncanny skill as a translator of the classical commitments voiced by EPR into the language of mathematics is remarkable, all the more so given that he was equally at home in the world of quantum theory. In this chapter, we shift our attention to the latter perspective. We start by reviewing some of the key experiments that shed light on the quantum prediction. We then consider in some detail the singular feature that distinguishes quantum theory from all other previous physical theories, namely, its use of the principle of superposition. Finally, we reexamine polarization, this time from the quantum perspective, for the sake of constructing the quantum prediction for Bell-type experiments. The concepts introduced here form the core of the vocabulary employed in the next chapter, which treats various philosophical interpretations of quantum entanglement. In particular, this chapter's discussion of photon polarization prepares the way for the next chapter's discussion of quantum theory's so-called "measurement problem."

In light of the increasingly sophisticated Bell experiments being done today, as well as the burgeoning literature on the meaning and possible applications of quantum entanglement to high-speed computing and global communications (see Brown 2000), it is clear now, almost five decades later, that what Bell uncovered was no mere curiosity of the quantum realm. Discoveries initially heralded as "society-changing" often appear less spectacular with time, but in Bell's case it was the reverse. At the beginning there was no heralding at all; Bell's 1964 paper was virtually ignored by the wider physics community for the better part of a decade. In the mid-1980s Leslie Ballentine researched the number of citations Bell's theorem had garnered in the scientific literature during the years following its publication. For the first nine years there were fewer than ten citations per year and in most years there were fewer than five (Ballentine's data are reproduced in Greenstein and Zajonc 1997, 107). The number began to increase in the mid-1970s, as word spread about the significance of Bell's work. Around the same time, the first experimental tests of a Bell-type inequality were performed.

1 EXPERIMENTAL EVIDENCE

The first scientists to take an interest in Bell's discovery needed to overcome two large hurdles before they could hope to perform tests of Bell's inequality in the laboratory. The first was that Bell's derivation of his inequality depended on the theoretically reasonable but experimentally unrealistic assumption that during some measurements the polarization filters on both sides of the apparatus would be set to exactly the same angle (see Figure 3.5). From an experimental point of view, the concern was that any slight misalignment of filters would cast doubt on the results (Aczel 2002, 147). The second hurdle, mentioned briefly in the previous chapter, was that neither real filters nor real detectors work like the ones in thought experiments. At the time, high-quality photon detectors registered only 10–20 percent of the incident photons. How was one supposed to know whether a photon that failed to register at the detector had been correctly absorbed by the filter, correctly transmitted by the filter but not detected, or incorrectly absorbed by the filter? On the other hand, how could one distinguish between photons that registered correctly at the detector and those that had been incorrectly transmitted by the filter?

The solution to both problems appeared in 1969 when John Clauser, Michael Horne, Abner Shimony, and Richard Holt (1969) published a new version of Bell's inequality. Known as the "CHSH inequality," it was the first of many "Bell-type" inequalities to come. Clauser et al. managed to avoid same-axis measurements by introducing a fourth measurement angle, which allowed two measurement angles to be chosen for one side and two for the other, each without any regard for the other. They then refashioned the ideal version of this new four-angle inequality into a more realistic expression that referred only to detection events. Experimentalists now had their first usable Bell-type inequality.

But where to turn for a source of entangled photons? Several decades earlier John Wheeler (1946, 219) had proposed that the polarizations of two photons emitted by a collapsing "atom" of positronium—an electron and positron caught in each other's electric fields—would be correlated. This scenario was actually realized in the laboratory several years later (Wu and Shaknov 1950), but no one at the time realized that the first pair of artificially entangled photons had been created. Bohm and Aharonov (1957) later pointed out the relevance of the Wu–Shaknov experiment, going on to suggest a possible experiment to determine whether the correlation remained or disappeared over large distances. They proposed, however, to measure the polarization of the two photons along the *same* axis, which would not have worked as a test of entanglement because both the quantum and classical approaches predict the same outcome in this case: perfect anti-correlation.

The first tests of the CHSH inequality involved photons separated by a distance of a few meters (the length of a standard optical bench). The first notable experiment, performed by Stuart Freedman and John

Clauser (1972), used excited calcium atoms as a source of photon pairs. The atoms' electrons were theorized to emit two entangled photons as they returned to their ground state from an excited state via a two-step process called an "atomic cascade." The electron took one of two possible routes back to the ground state, and because there was no way even in principle to tell which route it took the polarizations of the two photons it shed en route were expected to be entangled.[1] Most of the early experiments using atomic-cascade sources violated Bell's inequality by a wide margin. Two did not (Holt and Pipkin 1974; Faraci et al. 1974, 63–65), but their nonviolating results were never successfully reproduced. All of the early experiments, however, were vulnerable to the "locality" loophole discussed in the previous chapter because their filter orientations were set well ahead of each run.

In the early 1980s Alain Aspect, Jean Dalibard, and Gérard Roger performed a series of highly regarded experiments that produced the first violation of Bell's inequality with "dynamically determined" filter settings (1981, 1982a, 1982b). Aspect and his colleagues introduced a switch near the end of each wing of the experiment that effectively flipped the filter setting back and forth at a rate of roughly 100 million times per second. They detuned the two switches to ensure that the filter orientation in one wing would not always be the same as the orientation in the other wing. They also extended the distance between the detectors to 12 m, a feat that confirmed the presence of entanglement over a distance far greater than anyone had expected. Once again, the results agreed with the quantum prediction and violated the CHSH inequality by a wide margin.[2] In the mid-1980s, Leonard Mandel gave the new generation of experiments a boost when he surmised that the polarizations of the so-called "secondary photons" produced by an ultraviolet laser beam passing through a particular type of crystal—a process called "spontaneous parametric down conversion (SPDC)"—would be entangled (Ghosh and Mandel 1987). Ever since Mandel showed that the down-conversion process is easier to set up and more reliable than the atomic-cascade technique, SPDC has been the preferred source of photons for entanglement experiments.

Recent work in the laboratory has considerably strengthened the case for quantum entanglement by testing for its presence under increasingly stringent conditions. In one notable experiment, Wolfgang Tittel et al. (1998) used telephone fiber-optic cables to send entangled photons to detectors separated by a distance of more than 10 kilometers. The experiment was not designed to test Bell's inequality. Instead, the group wanted to answer the question of whether entangled states could be maintained over large distances. Their results put to rest any lingering doubts about whether entanglement qualified as a "macroscopic" phenomenon. More recently, Rupert Ursin et al. (2007) have demonstrated the persistence of entanglement over 144 kilometers, a distance sufficiently large to consider using entanglement-based encryption techniques for satellite communications.

In another highly ambitious experiment, Gregor Weihs and colleagues (1998) set out to eliminate the problems with Aspect's earlier experiments.

To prevent any preliminary local signaling by the filters and yet have their orientations established before the photons arrived, the group needed to set the filters' orientations within a few billionths of a second after the photons had been emitted by the source. Weihs et al. went to great lengths to ensure that no luminal or subluminal communication could occur between different parts of the apparatus. They used a quantum-based random number generator to establish the filter orientations; they reduced the duration of the entire measurement process (from the generation of the random number to the registration of the photon) to less than one-tenth of the time needed to guarantee signal isolation between the two wings; they compared data from each wing only afterward via a computer rather than during the experiment; and they synchronized the atomic clocks used to time-tag each measurement only once before the experiment. With all of these precautions in place, they saw the clearest violation of Bell's inequality yet.[3]

Mary Rowe et al. (2001) achieved another important milestone when they overcame for the first time the other key difficulty plaguing all prior experiments, including the Weihs experiment: the so-called "detection-efficiency loophole" (see 164 n. 26). This group entangled not photons but beryllium ions, which are relatively large and easy to detect. This allowed them to achieve 100 percent detection efficiency and close the detection loophole for the first time. However, the relative proximity of the ions to one another at the time of measurement (around 3 millionths of a meter) prevented the group from also closing the locality loophole. A single "loophole-free" experiment has yet to be performed. Some have questioned whether such an experiment will ever be possible (Sandu Popescu, personal communication; Barrett et al. 2002), but several different approaches have been suggested (e.g., Simon and Irvine 2003; García-Patrón et al. 2004). Although it would be reassuring to have a loophole-free test, few physicists today think that the observed correlation results are due simply to local, loophole-exploiting causes masquerading as entanglement.

Most recently, Daniel Salart and colleagues (2008) have looked into the issue of entanglement's speed. Although generally discounted as an explanation, it is at least theoretically possible that the correlations could be due to superluminal communication, which is not strictly prohibited by Einstein's special theory of relativity.[4] Even though the dynamics of nature appear to prevent us from communicating superluminally, perhaps the world itself can access superluminal channels of influence. Salart et al. established a lower bound on the speed of any signal that could produce results consistent with the quantum prediction. By minimizing and then carefully measuring the time between the measurement events in each wing of the experiment, they determined that the speed of any hypothetical influence would need to be at least ten thousand times greater than the speed of light in all cases and roughly one hundred thousand times greater on average.

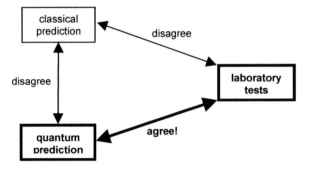

Figure 4.1   Laboratory results agree with quantum prediction.

2   SUPERPOSITION IN QUANTUM THEORY

Quantum theory was constructed in the early decades of the twentieth century in an effort to solve a number of outstanding problems about microphysical (i.e., atomic-scale) processes, such as the stability of electron orbits around atomic nuclei and the curious spectral patterns of light given off by electrically excited gases. The theory has been enormously successful, generating innumerable well-tested predictions, and it has become the basic framework for much of modern physics. But it has also generated unending debate. As with Newtonian physics, quantum mechanics requires interpretation. Well defined rules link its mathematical structure (i.e., "formalism") to experimental results, but the formalism cannot say what the theory *means*. Only interpreters can do this (cf. Howard 1989, 225 n. 1). Broad interpretive agreement followed the rise of Newtonian physics, namely, that nature was a closed, deterministic system, but no such consensus has appeared in the case of quantum theory. In fact, the quantum formalism has led to just the opposite: a variety of metaphysically conflicting interpretations (see Chapter 5).

At the heart of the debate lies a mathematically simple but ontologically ambiguous idea: the "principle of superposition." Paul Dirac, one of the founders of quantum theory, called quantum theory's use of superposition a "drastic" departure from the world of classical physics (1958, 4). Richard Feynman called it quantum theory's "*only* mystery" (Feynman, Leighton, and sands 1965, chap. 1; italics in original). Entanglement, it turns out, is a special kind of superposition involving multiple objects. In order to see why entanglement calls into question the classical worldview, we need to consider first what is so *un*classical about quantum theory's use of the superposition principle. This will allow us to consider briefly several of the more prominent interpretations—and even revisions—of the basic theory that have been put forward. And once we have surveyed the broader interpretive landscape, we will be in a good position to understand the various interpretations of quantum entanglement to be discussed in the next chapter.

What does it mean to say that superposition implies the nonclassical, "wave-like" nature of quantum particles (photons, electrons, protons, and the like)? To be concrete, let's return to the photon. We will focus on the polarization values T_v and S_v of the property "vertical polarization" (this is still using classical language, *pace* the challenge of imaging particles having polarization) along with the respective measurement outcomes T and S. The classical physicist begins with the simple observation that these two outcomes are, in fact, mutually exclusive—one never finds both outcomes at the same time in a single measurement. She then infers, straightforwardly, that the corresponding values of the property must be mutually exclusive as well, in the sense that the vertical polarization state of any photon must always be linked to one value or the other—T_v or S_v—and nothing else. In the previous chapter we referred to this inference as *faithful measurement*.

The quantum physicist rejects this inference. From the standard[5] interpretive point of view, any simple sum[6] of two or more classical states is itself regarded as a physically possible state, even though the values that comprise the sum are associated with mutually exclusive outcomes. For example, T_v and S_v are perfectly acceptable classical states—in the sense that each ascribes a particular value to a particular property. They are also mutually exclusive states; no combination of these two states, such as $T_v + S_v$, makes any sense from the classical point of view. According to quantum theory, however, classically allowable states can be added together or "superimposed" in this manner (see Figure 4.2) to form a new, distinct—and physically possible—state. Quantum "states," then, include states in which objects carry some particular property only indefinitely (according to the standard interpretation, at least). Talk of the "wave-like" nature of a quantum state refers to situations in which some property (e.g., position) can be associated with an infinite rather than finite number of values. The "continuous" nature of the wave description captures this aspect of the state, and its "height" at any given point corresponds, in the case of position, to the likelihood of finding the object at some particular place.

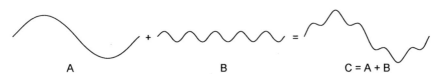

A   B   C = A + B

Figure 4.2 The principle of superposition. Classically, the superposition of any two waves A and B—whether they be ripples on the surface of a pond, undulations in an electromagnetic field, or pulsations through a slinky—yields a third wave C. According to quantum theory, any two physical states A and B—even those associated with mutually exclusive outcomes, like "going through this door" and "going through that door," can be added together or superimposed to form a new, physically realizable state C = A + B. The odd thing about such states, from the classical point of view, is that they do not lead to just one outcome when measured. When an object in state C is measured, two outcomes are possible—those associated with states A and B. Because state C is associated with two possible outcomes rather than one, it is said to be an "indefinite" state with regard to the property in question (e.g., vertical polarization).

As Roy Glauber put it in his Nobel Lecture, thinking about the states of particles like photons in this way would seem to lead to the idea that such particles "interpenetrate one another like waves on the surface of a pond" (2007, 6). Mathematically speaking, the amplitudes of the waves add together (as in Figure 4.2). In particular, where the waves' amplitudes are both positive or both negative they reinforce one another and are said to "interfere constructively." Where their amplitudes are oppositely signed, they cancel one another out and are said to "interfere destructively." A single photon, for example, can exist in the T_V state, but it can also exist in the $S_V - T_V$ (superposition) state. Taking this one step further, it can also exist in a superposition of these two states: $T_V + (S_V - T_V)$. In this case, however, the first and last terms interfere "destructively," which is to say that a photon in the state $T_V + S_V - T_V$ is simply in the S_V state. Here is the puzzling feature of superposition: The superposition of two states, both of which contain the basic state T_V, add together to yield a state which *lacks* the T_V state. A photon prepared in $T_V + (S_V - T_V)$ state will, oddly enough, never yield T_V as an outcome because these two components of its overall state interfere "destructively" so as to make the state equivalent to S_V. In the language of waves, although it is not strictly applicable in this case, one would say that the height of the quantum wavefunction at the point corresponding to the outcome "transmit" is zero.

Here are two sets of superpositions (derived and explained in detail in the following) that describe the reciprocal relations between the two possible polarization outcomes (S = stop and T = transmit) for two different measurements, one along the vertical axis and the other along some arbitrary axis α. Each of the two polarizations for α can be expressed as a superposition of the two vertical polarizations:

$$T_\alpha = aT_V + bS_V \qquad\qquad 4.1$$

$$S_\alpha = -bT_V + aS_V, \qquad\qquad 4.2$$

where $a = \sin(\alpha)$ and $b = \cos(\alpha)$ and α is just the angle between the two measurement axes. The leading variables a and b are simply numbers that give the "strength" of each basic state within the superposition. The relationship runs the other way as well, so that each of the two vertical polarizations can be expressed as a superposition of the two α polarizations:

$$T_V = aT_\alpha - bS_\alpha, \qquad\qquad 4.3$$

$$S_V = bT_\alpha + aS_\alpha. \qquad\qquad 4.4$$

These basic relations can help us understand the controversy regarding the meaning of superposition. The standard view is that a photon in the T_α-polarization state is in a "definite" α-polarization state in the sense that measuring this photon's polarization along α will always lead to the same outcome for such a photon (T, or +1). On the other hand, and because of Equation 4.1, the same photon is said to be in neither the T_v-polarization state nor the S_v-polarization state. Instead, it is said to be in a nonclassical "indefinite" combination or superposition of these two states. In fact, all four of the expressions on the right-hand side of Equations 4.1–4.4 are nonclassical superpositions, whereas each of the simpler expressions on the left-hand side is classically interpretable as a definite state.

The point to take away from these otherwise simple relations is that they thwart any attempt to say generally whether or not a particular photon is in a definite state. According to the standard view, this kind of judgment can only be made relative to a specific filter orientation. Equation 4.3, for example, implies that although a photon with the T_v-polarization property is in a definite polarization state relative to v-measurement, the same photon is in a superposition of basic polarization states relative to α-measurement (i.e., the "$aT_\alpha - bS_\alpha$" state). More generally, the standard view insists that a photon in a basic polarization state relative to some type of measurement is necessarily in a superposition of basic polarization states relative to nearly every other possible polarization measurement. One definite property necessarily implies many other indefinite properties. This is precisely what it means to say that the standard interpretation of quantum theory abandons the classical principle of "property definiteness" and embraces the idea of "property indefiniteness" through the principle of superposition.

Let me offer one final comment about superpositions. Despite their proven track record when it comes to generating accurate predictions, they are not something physicists ever expect to observe directly in the laboratory. When a physicist sets out to measure the v-polarization state of a photon she regards to be in the "$T_v + S_v$"-polarization state, she only ever expects to find a T_v-polarized photon *or* an S_v-polarized photon. In other words, she never expects to observe directly anything but states that are "basic" to this type of measurement. Readers should keep this important point in mind in the following, where the phrases "basic state" and "observed outcome" will be used interchangeably. In this regard, but in this regard only, the quantum physicist's expectations are the same as those of the classical physicist. The quantum physicist's approach differs in that she goes on to account for the randomness of outcomes by inferring the existence of superpositions or indefinite states from the relative abundance of various basic states as they appear (randomly) in measurements of many photons all prepared in the same way. Nonetheless, the absence of superpositions in observed outcomes has not in general led quantum physicists to view these states as mere figments of the imagination. On the contrary, the indispensible role of superpositions in explaining the apparent randomness of individual outcomes and in aligning the predictions of quantum theory with experimental results has led physicists to take superpositions seriously as meaningful descriptions of physical states, not just with regard to polarization but with regard to the states of microphysical objects in general, even as they confess not to know exactly what a superposition is.

From the classical point of view, the notion that a physical object can exist in a superposition but never be observed *qua* superposition is contentious to say the least. After all, making a measurement is the classical *sine qua non* of determining an object's state. That's what measurements do—they measure states! If I pick up a cup of water and find it frozen solid, I presume that my observation straightforwardly reveals the state of the water just before I looked. Likewise, if I measure a photon and find that it is T_V-polarized, my inner classical physicist presumes that this outcome obtains because the photon was in fact T_V-polarized just before I measured it. From the classical point of view, claiming that the outcome might have occurred because the photon was in a "$T_V + S_V$"-polarization state makes no more sense than claiming that the water in the cup might have been simultaneously frozen solid and boiling just before I looked. How photons and other quantum particles go from being in superpositions like "$T_V + S_V$" to basic states like "T_V"—and whether they really do make this transition at all—is a good example of what is commonly referred to as quantum theory's "measurement problem" (von Neumann 1955). Quantum theory, with its uncanny application of the superposition principle, forces us to rethink the formerly "obvious" idea that a measurement always gives us a clear picture of an object's state. The apparent fact that physical objects sometimes exist in states that hide themselves from our direct gaze, at least according to the standard interpretation, makes the determination of physical states a much more subtle task than classical physicists had imagined.

It is important here not to be misled into thinking that the problem has to do simply with the fact that quantum theory uses the concept of superposition. No physicist trained in the classical era would have denied the usefulness of this concept for talking about the wave-like behavior of large collections of particles. The surface of a lake, the vibrating strings of a violin, air pulsing back and forth inside a trumpet—these and many other everyday collections of particles can and often do exist in complex superpositions of their basic collective states (e.g., a violin string can vibrate in its fundamental and upper harmonic modes simultaneously; the relative strength of these modes gives the sound of an individual violin its particular character). But classical physicists regarded such wave-like behaviors as manifestations of the more fundamental particulate nature of the objects' micro-constituents. Although it is true that a violin string can sustain a superposition of more than one wave mode at the same time, the classical physicist would have regarded this as a matter of the string's micro-constituents behaving as they always do—moving along with well-defined positions and velocities that give rise in this case to the global wave-like behavior of the string. The real problem with the standard interpretation of quantum theory, from the classical perspective, is that it understands superpositions to describe the behavior of *individual* particles and not just the collective behavior of particles in large groups.

But even allowing the necessity of invoking wave-like combinations of states to account for the behavior of photons, are we not still entitled to presume that a given photon is always actually in some particular (i.e., definite, basic) state? To put the question more concretely, even if we allow that the superposition of polarization states given on the right-hand side of Equation 4.1 accurately predicts what happens when we measure the vertical polarization of an ensemble of photons polarized in some other direction, must not each individual photon within the collection actually have been specifically and definitely determined to be transmitted or stopped by a vertical polarizer before we measured it? The standard quantum account of polarization says no. Instead, it claims that a T_α-polarized photon, i.e., a photon in the state described by the superposition of Equation 4.1, is not in a definite v-polarization state until it undergoes a v-polarization measurement and is forced to assume a definite v-polarization state. According to the standard interpretation, then, measurements of quantum objects don't just reveal their states, they *change* them. Such an odd-sounding claim might lead one to seek haven in the apparently reasonable position that superposition merely reflects our ignorance of the world's actual, definite state of affairs. Einstein held firmly to this view—what we have been calling "property definiteness"[7]—just as he insisted upon the ideas of state separability and cause locality (Einstein 1948, reprinted in Born 1971, 168–73; Einstein 1970; see also Born 1971, 164; Howard 1985). But as the development of Bell's argument in Chapter 3 made clear, there is a problem with affirming property definiteness along with state separability and cause locality. Their conjunction leads to a demonstrably incorrect prediction in the form of a Bell-type inequality. David Bohm's subsequent formulation of a definite-property version of quantum theory (1952) showed that it is possible to recover property definiteness and maintain empirical accuracy, but Bohm's alternative formulation offers no return to the classical perspective; it is entangled in its own way (more on this in the following chapter).

One of the most puzzling aspects of quantum theory is that, despite its successful and ubiquitous use of the superposition principle to account for atomic and subatomic behaviors, one never finds things like a pot of water—which, after all, is made of quantum particles—behaving as though it were both frozen and boiling at the same time even though such behavior is possible according to quantum theory. What we do know now is that it is wrong to interpret superposition merely as a reflection of our ignorance of the otherwise definite, separable, and locality-abiding state of the world. If superposition were a matter of ignorance, and if the classical worldview held on all three points, then experiments would have confirmed Bell's inequality. But they didn't.

Quantum theory's use of superposition is puzzling for another reason as well. The mathematical formalism of the theory makes no distinction between microphysical and macrophysical objects. There is no reason, according to the formalism, why the basic principle of superposition cannot be applied to any physical object, regardless of its size or complexity. According to quantum theory, we can treat both a photon and the device used to measure its polarization as capable of existing in superposition states. If we ask what state the combined particle/apparatus system is in after they interact, quantum theory forces us to say that the apparatus has become entangled with the particle (and vice versa) so that the apparatus itself has now become "infected with superposition" (Redhead 1987, 54). The difficulty with this view, of course, is that macrophysical objects don't typically (!) appear to have indefinite properties. Property definiteness is a bedrock presupposition of classical physics precisely because it accords well with everyday experience. So how can a world made out of indefinite superpositions at the atomic and subatomic scales manifest itself to us in a seemingly definite way? Because we do not experience superpositions, we need to explain within the quantum framework why a classical measuring apparatus ever registers a definite result. This nexus of issues is referred to as quantum theory's "measurement problem" because the issue is typified in the dynamics of measurement. This problem is almost exclusively responsible for the variety of interpretations of quantum theory that have been proposed since its inception; all of the major interpretations beyond the standard one are basically attempts to solve it, or at least to provide a clear mechanism by which superpositions collapse onto definite states. We will return to this issue in the next chapter, but for now we have a sufficient sense of how quantum theory uses superposition to return to the concept of polarization.

3 PHOTON POLARIZATION, AGAIN

It is precisely the principle of superposition that allows quantum theory to describe the polarization of photons in such a way that its predictions match experimental results. One can get a feeling for the quantum approach to polarization, and even see why it leads to a violation of Bell's inequality, with the help of only a few basic tools of algebra, the barest rudiments of trigonometry, and a readily intuited concept that combines both, namely, the "vector" (a helpful introduction to vectors, superposition, and quantum theory in general can be found in Maudlin 2002, 243–267).

A vector is a mathematical object with both magnitude *and* direction (regular workaday numbers, or "scalars," have magnitude or "scale," but no direction). If I were driving from Des Moines to Minneapolis at 65 miles per hour, my speed—a scalar—would be just that: 65 miles per hour. On the other hand, my velocity—a vector—would be 65 miles per hour *in a northerly direction*. Imagine an arrow drawn on a sheet of graph paper pointing up 65 squares; this is one way of representing the vector. Vectors can easily be "added" geometrically by juxtaposing their arrow representations in head-to-tail fashion. A new arrow drawn from the tail of the first to the head of the last represents their "vector sum." An example of this, the meaning of which will be explained in the following, appears in the inset of 4.4b where the vector S_a is expressed in terms of the following scaled sum of vectors T_V and S_V:

$$S_a = -\cos(\alpha)T_V + \sin(\alpha)S_V.$$

The visual representation of vectors as arrows is helpful for strengthening intuitions, but including vectors in algebraic expressions requires a different approach called "vector notation." In this case (as in the preceding equation) a scalar, typically referred to as the vector's "coefficient," indicates the vector's size or magnitude; an additional symbol indicates its direction. For example, "65↑" denotes a vector pointing 65 units up (e.g., driving 65 miles per hour in a northerly direction). A vector's size can be changed by multiplying or "scaling" its coefficient by some new number (e.g., $-2 \times 65↑ = -130↑$); changing a vector's mathematical sign simply reverses its direction (e.g., $-130↑ = 130↓$). Vectors describing quantum states often have simple trigonometric functions as their coefficients, e.g., "$\cos(\alpha)↑$ or $-\cos(\alpha)T_V$," but no matter how complicated these coefficient functions are, they always stand for a single number, which represents the vector's magnitude. For example, if $\alpha = 180°$, then it is clear from Figure 4.3 that "$\cos(\alpha)↑$" reduces to "$-1↑$" or just "$-↑$" (or alternatively, "↓").

In quantum theory, we will associate each of the two possible outcomes for any polarization measurement (transmit and stop) with a vector. We will call these two vectors the "basic polarization vectors" of a given filter orientation; together they form the complete "set" of basic polarization vectors for that orientation in the sense that they correspond to all of the possible measurement outcomes for a given orientation. In the particular case of a vertically oriented filter, the two basic polarization vectors will be represented as "T_V" and "S_V" (where the subscript v stands for "vertical" or $\alpha = 90°$). The *length* of a basic polarization vector is always 1, and it is usually omitted from the description (e.g., $1T_V = T_V$). The portion of the overall symbol indicating magnitude is normally given in regular type; whereas the portion indicating direction is normally given in boldface (the use of boldface in the previous chapter anticipated this).

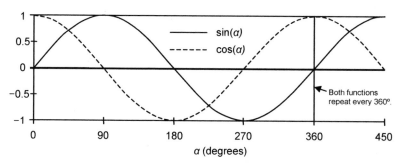

Figure 4.3   Sine and cosine. The values of the periodic functions sine and cosine for angles ranging from 0° to 450°.

The most important feature of basic vectors is that their mathematical relationship embodies the mutual exclusivity of basic outcomes. The fact that a photon prepared in the state T_V will always be transmitted and never stopped by a vertically oriented polarization filter is represented in the world of vectors by the fact that basic vectors "contain" nothing of each other. In more technical terminology, they are "orthogonal" to one another. In standard two-dimensional Cartesian geometry this simply means that they are at right angles to each other. Thus if we represent the basis vector for the state T_V by an arrow pointing one unit up (90°), we will want to represent the state S_V by an arrow pointing one unit to the right or left; we choose right by convention (0°). Figure 4.4a uses unit-length arrows to depict the basic photon polarization vectors T_V and S_V, which together express the polarization "basis" corresponding to measurement with a vertically oriented filter. An arrow pointing one unit up "contains" no element of an arrow pointing one unit to the right in the sense that it points not at all in the horizontal direction. The same is true of the horizontal arrow, of course. It points not at all in the vertical direction. The benefit of basic vectors is that any other vector in the same space can be represented as a vector sum of these two basic vectors or any other pair (there are infinitely many), as long as each is appropriately scaled. The inset in Figure 4.4b shows S_a as the scaled sum of the two basic vectors T_V and S_V.

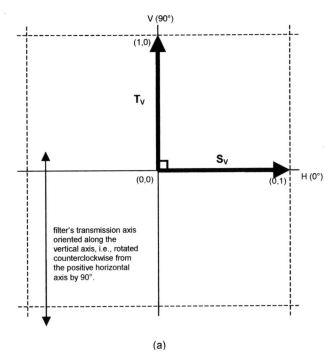

(a)

Figure 4.4a The two basic polarization vectors T_V and S_V associated with a vertically oriented filter ($\alpha = 90°$). The two vectors are orthogonal and hence contain no element of the other. This corresponds to the physical exclusivity of the two measurement outcomes they represent: "transmit" and "stop."

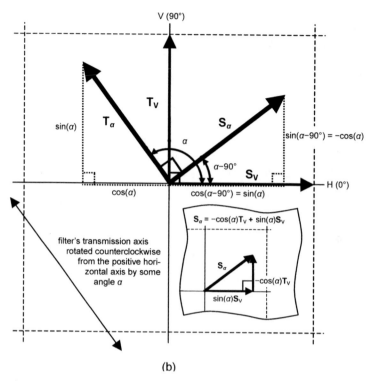

(b)

Figure 4.4b The two basic polarization vectors T_α and S_α associated with an α-oriented filter superimposed on Figure 4.4a. The magnitude of all basic polarization vectors—$|T_V|$, $|S_V|$, $|T_\alpha|$, $|S_\alpha|$—is equal to 1. Either basic vector in either pair can be represented as a scaled sum of the other pair. Inset: As an example, the basic vector S_α can be represented geometrically and algebraically as the sum or superposition of the vectors T_V and S_V through the expression $S_\alpha = -\cos(\alpha)T_V + \sin(\alpha)S_V$.

In Chapter 3, we developed a rudimentary classical model of polarization by assuming that a single photon simultaneously has definite polarization values associated with different polarization properties or measurements (e.g., "$T_{90°}$, $T_{45°}$"). We did this in order to maintain the principle of property definiteness. We did not attempt to say how the values associated with different properties (i.e., different measurement angles) might relate to one another mathematically or theoretically. Figure 4.4b, which superimposes onto Figure 4.4a a second set of basic vectors, T_α and S_α (the basic polarization vectors associated with an α-oriented filter), provides a visual representation of how this relationship is expressed in the quantum formalism, according to which there is a simple though powerful relation between any two basic sets of polarization properties. Recall that the length of the side opposite one of the acute angles of a right triangle divided by the length of its longest side, the hypotenuse, is equal to that angle's sine, and that the length of the side adjacent to the angle divided by the length of the hypotenuse is equal to the angle's cosine. From these two definitions one can see that the vertical and horizontal distances covered by S_a are the sine and cosine of $\alpha - 90°$, respectively:

vertical distance covered by $S_a = \sin(\alpha - 90°)$,

horizontal distance covered by $S_a = \cos(\alpha - 90°)$.

In this form, these expressions are helpful for relating the vector S_a to the vectors T_V and S_V.

Figure 4.4b reveals that the vertical distance covered by S_a is equal to the "scaled" vector $\sin(\alpha - 90°)T_V$. This vector lies in the direction of T_V and is commonly referred to as the "projection" of S_a onto T_V (imagine the length of the "shadow" cast by S_a onto the vertical axis by shining a light onto this vector from the right). Likewise, the horizontal distance covered by S_a is equal to the scaled vector $\cos(\alpha - 90°)S_V$, which is the projection of S_a onto S_V (now imagine a light shining from above). We can express the vector S_a as the sum of these two scaled vectors:

$$S_a = \sin(\alpha - 90°)T_V + \cos(\alpha - 90°)S_V.$$

We can do the same with T_a. Because the angle α is defined in terms of this basic vector, we can express its projections directly in terms of α: The projection of T_a onto the vertical axis is given by $\sin(\alpha)$, whereas the projection of T_a onto the horizontal axis is given by $\cos(\alpha)$.[8] Combining all of these observations, and making use of the basic trigonometric identities

$$\cos(\alpha - 90°) = \sin(\alpha)$$
$$\sin(\alpha - 90°) = -\cos(\alpha)$$

(which, again, the reader can confirm visually by examining Figure 4.3), we can express each of the two basic α-vectors in terms of α and the two basic v-vectors:

$$T_a = \sin(\alpha)T_V + \cos(\alpha)S_V, \qquad\qquad 4.5$$

$$S_a = -\cos(\alpha)T_V + \sin(\alpha)S_V. \qquad\qquad 4.6$$

And with a little algebraic manipulation (which I leave to the curious reader), these two equations can be rearranged to give each of the two basic v-vectors in terms of the two basic α-vectors:

$$T_V = \sin(\alpha)T_a - \cos(\alpha)S_a, \qquad\qquad 4.7$$

$$S_V = \cos(\alpha)T_a + \sin(\alpha)S_a. \qquad\qquad 4.8$$

Notice that each of the equations reduces to a simple identity when $\alpha = 90°$. Mathematically speaking, Equations 4.5–4.8 describe quantum theory's view of the relationship between different sets of basic polarization vectors. They imply that the different sets of basic polarization vectors we associated in the previous chapter with separate properties are not independent, but are related to one another through the angle separating their respective filter orientations. In general terms, any polarization vector (or state) can be redescribed as a *superposition* of some particular set of basic polarization vectors.

These relationships allow us to partially reclaim the classical idea that light can be manufactured to have one particular polarization state. The necessary complication, from the quantum mechanical point of view, is that no claim of this type can be proffered as *the only* possible description of the light's polarization because its polarization state can be expressed in an infinite variety of ways, each of which gives the state (of each individual photon) in terms of a specific type or angle of measurement.

4 MAKING PREDICTIONS

The superposition relations given in Equations 4.5–4.8 describe the mathematical relations between different sets of basic polarization vectors, but they also lead to quantitative predictions for polarization measurements. When the polarization state of a particular photon must be expressed as a superposition with reference to an intended measurement, quantum theory refuses to make an absolute prediction about which particular outcome will occur (e.g., quantum theory refuses to say in this case, "When one measures the α-polarization of photon A, one will most certainly find T."). Instead, the theory regards each of the outcomes present within a superposition as "possible" and identifies the likelihood of each outcome occurring in the form of a probability (e.g., it might allow one to say, "When I measure the α-polarization of photon A, there is a 75 percent chance I will find T and a 25 percent chance I will find S."). This important principle was first enunciated by Max Born, one of the early developers of quantum theory, and is thus part of what is eponymously referred to as the "Born interpretation" of superposition.[9]

One relevant point in this regard is that the probabilities associated with different outcomes within a superposition must always sum to one when all of the possible outcomes are taken into account (by definition, the probability of finding *some* outcome during a measurement event is 100 percent). Notice that the coefficients of the two vectors on the right-hand side of Equation 4.5, curiously, do not sum to one. However, the squares of these coefficients do, as do the squares of the coefficients in Equations 4.6, 4.7, and 4.8—all in accord with the basic trigonometric identity $\sin^2(\theta) + \cos^2(\theta) = 1$. In light of this observation, a reasonable guess would be that in the quantum formalism the square of a basic vector's coefficient gives the probability of finding the corresponding value in measurement. In fact, this is the approach adopted by quantum theory, not just for polarization but for *all* dynamical quantities. It is a consequence of the theory's use of orthogonal unit vectors and their sums (i.e., superpositions) to represent all physical properties.[10]

For our first exercise in using the formalism to make a prediction, we will find the probability of a T_a-polarized photon being transmitted by a v-oriented filter. In the language of quantum theory, we need to determine the probability of measuring the v-polarization of a photon we know to be T_a-polarized and finding the photon is transmitted. From Equation 4.5 we know that a T_a-polarized photon is equivalent to a photon in the following superposition (i.e., indefinite) state: "$\sin(\alpha)T_V + \cos(\alpha)S_V$." To determine the probability of the photon being transmitted (finding T_V as the outcome), we square the coefficient associated with this basic polarization value as it appears within the overall superposition to get $P(T_V) = \sin^2(\alpha)$. An initially T_a-polarized photon will thus behave probabilistically when measured along the v-axis; it will be transmitted with a probability of $\sin^2(\alpha)$ and stopped with a probability of $\cos^2(\alpha)$.

This example focuses on the behavior of an individual photon, whereas EPR–Bell experiments generate data for jointly produced *pairs* of photons. How does quantum theory generate predictions for joint behaviors? In short, it does so by multiplying together individual coefficients to produce "joint" coefficients, which then characterize the overall "joint" vector of the system (i.e., the overall joint-polarization state of the two photons). Probabilities are then obtained in the usual way, through squaring.[11] As in Chapter 3, we will indicate joint states with the symbol $\otimes$, the general properties of which again need not concern us; we will continue to treat it as an analogue of algebraic multiplication for states.

As a second exercise, let's consider the following situation. The joint-state vector "$\psi_{AB} = T_V \otimes S_V$" represents a joint-polarization state in which a T_V-polarized photon flies toward the left filter and an S_V-polarized photon flies toward the right filter (I have used ψ_{AB} to denote the overall joint state, according to convention). Recall that a simple joint state such as this one is called a "product" state because it can be expressed as a single product of individual photon polarization vectors under some joint measurement scenario, i.e., a product of individual vectors involving only one $\otimes$. What is the probability of measuring the vv-polarization of a $T_V \otimes S_V$-polarized pair of photons and finding that the left photon is transmitted but that the right photon is stopped? To calculate the probability of this outcome, we multiply together the individual coefficients and square the result. In order to prevent any confusion over whether a particular coefficient is "owned" by an individual or joint vector, I will place joint coefficients outside curly braces—a process I will call "contraction." Writing down the coefficient of each individual vector explicitly, the contraction of "$T_V \otimes S_V$" looks like this:

$$\mathbf{T_V} \otimes \mathbf{S_V} = (1)\mathbf{T_V} \otimes (1)\mathbf{S_V} \xrightarrow[\text{contraction}]{} (1 \times 1)\{\mathbf{T_V} \otimes \mathbf{S_V}\} = (1)\{\mathbf{T_V} \otimes \mathbf{S_V}\} = \{\mathbf{T_V} \otimes \mathbf{S_V}\}.$$

As always, we square the coefficient to find the probability. In this case, the result is trivial. When we measure the vv-polarization state of an initially $\mathbf{T_V} \otimes \mathbf{S_V}$-polarized pair of photons, quantum theory predicts that the probability of finding the left photon transmitted and the right photon stopped is just $1^2 = 1$, or 100 percent. In other words, the final joint state of the photon pair is guaranteed to be the same as its initial joint state. The fact that this simple product state reproduces itself with certainty in the case of a vv-measurement confirms that it is one of the basic joint-polarization states of the vv-measurement scenario. The other three basic joint states of this measurement scenario are $\mathbf{T_V} \otimes \mathbf{T_V}$, $\mathbf{S_V} \otimes \mathbf{T_V}$, and $\mathbf{S_V} \otimes \mathbf{S_V}$ (according to quantum theory, the outcome of any type of measurement can be known with certainty only when the initial state can be expressed in terms of *one and only one* of that measurement's basic states, i.e., not in terms of a superposition of basic states).

Now let's consider a nontrivial problem. What is the probability of obtaining the same joint outcome $(\mathbf{T_V} \otimes \mathbf{S_V})$ when we measure the vv-polarization state of a $\mathbf{T_\alpha} \otimes \mathbf{S_\alpha}$-polarized pair of photons? This problem is nontrivial because even though the initial state $\mathbf{T_\alpha} \otimes \mathbf{S_\alpha}$ is a simple product state, it is not a basic joint state of vv-measurement. It will help to divide the overall procedure for calculating the joint probability into five steps.

Step 1. Use Equation 4.3 to rewrite $\mathbf{T_\alpha} \otimes \mathbf{S_\alpha}$ in terms of the joint measurement we actually intend to perform (a vv-measurement):

$$\mathbf{T_\alpha} \otimes \mathbf{S_\alpha} = \left[\sin(\alpha)\mathbf{T_V} + \cos(\alpha)\mathbf{S_V}\right] \otimes \left[-\cos(\alpha)\mathbf{T_V} + \sin(\alpha)\mathbf{S_V}\right].$$

Step 2. Eliminate all superpositions "internal" to each photon's state by expanding the expressions around each product symbol $\otimes$ as we did in Chapter 3, i.e., by multiplying out the sums on either side of each $\otimes$ in the standard algebraic manner

$$(a + b) \times (c + d) = (a \times c) + (a \times d) + (b \times c) + (b \times d).$$

In the present case, there are two internal superpositions inside a single product state. Thus we obtain:

$$\begin{aligned}
\mathbf{T_\alpha} \otimes \mathbf{S_\alpha} = &\left(\sin(\alpha)\mathbf{T_V} \otimes -\cos(\alpha)\mathbf{T_V}\right) \\
&+ \left(\sin(\alpha)\mathbf{T_V} \otimes \sin(\alpha)\mathbf{S_V}\right) \\
&+ \left(\cos(\alpha)\mathbf{S_V} \otimes -\cos(\alpha)\mathbf{T_V}\right) \\
&+ \left(\cos(\alpha)\mathbf{S_V} \otimes \sin(\alpha)\mathbf{S_V}\right).
\end{aligned}$$

This step is important because only basic joint states, i.e., states with no internal superpositions, can be directly observed in measurement. Note that each of the four basic joint states of the vv-measurement scenario now appears within the description of the overall joint state. This means that any of the basic vv-states might turn up if we were to measure the vv-polarization of this pair of photons.

Step 3. Convert all individual coefficients to joint coefficients by contraction:

$$
\begin{aligned}
\mathbf{T}_a \otimes \mathbf{S}_a = {} & -\sin(\alpha)\cos(\alpha)\{\mathbf{T}_V \otimes \mathbf{T}_V\} \\
& + \sin^2(\alpha)\{\mathbf{T}_V \otimes \mathbf{S}_V\} \\
& - \cos^2(\alpha)\{\mathbf{S}_V \otimes \mathbf{T}_V\} \\
& + \cos(\alpha)\sin(\alpha)\{\mathbf{S}_V \otimes \mathbf{S}_V\}.
\end{aligned}
$$

Step 4. Group joint coefficients according to basic joint vectors (we can skip this step here because each basic joint vector appears only once in the expression).

Step 5. Square the coefficient associated with each basic joint vector to obtain the quantum mechanical probability, P_{QM}, of each joint outcome:

$$
\begin{aligned}
P_{QM}(\mathbf{T}_V \otimes \mathbf{T}_V) &= \sin^2(\alpha)\cos^2(\alpha) \\
P_{QM}(\mathbf{T}_V \otimes \mathbf{S}_V) &= \sin^4(\alpha) \\
P_{QM}(\mathbf{S}_V \otimes \mathbf{T}_V) &= \cos^4(\alpha) \\
P_{QM}(\mathbf{S}_V \otimes \mathbf{S}_V) &= \cos^2(\alpha)\sin^2(\alpha).
\end{aligned}
\qquad 4.9
$$

A little algebra (which again I leave to the reader) shows that these four probabilities still sum to one, i.e., to 100 percent, as they must if they are to represent all possible outcomes of a vv-measurement performed on a single pair of photons in this particular joint state. In light of these results, we can now give the quantum mechanical prediction: When performing a vv-measurement on a $\mathbf{T}_a \otimes \mathbf{S}_a$-polarized pair of photons, the probability of finding the outcome $\mathbf{T}_V \otimes \mathbf{S}_V$ is $\sin^4(\alpha)$.

Had we wanted to calculate the probability of the left photon being transmitted and the right photon being stopped under some other joint measurement scenario, such as the v-α scenario ($\mathbf{T}_V \otimes \mathbf{S}_a$) or the α-v scenario ($\mathbf{T}_a \otimes \mathbf{S}_V$), we would have begun with the single relevant substitution in each case and then proceeded as before. The following list summarizes the five basic steps for calculating the probability of any joint-polarization outcome:

1. *Express the overall state exclusively in terms of the polarization measurement to be performed on each photon.* Express all polarization vectors exclusively in terms of the types of measurement to be performed, substituting Equations 4.5–4.8 as necessary.
2. *Expand the joint vectors.* Multiply out all joint vectors with "internal" superpositions to eliminate them because only basic joint states are directly observable in measurement.
3. *Find joint coefficients by contracting individual coefficients.* Multiply together the individual coefficients of each resulting basic joint vector, then place them outside curly braces.
4. *Group joint coefficients.* If the same basic joint vector appears more than once within an overall expression, group together (by adding) all joint coefficients for that vector.
5. *Square joint coefficients.* Square the resulting joint coefficients to obtain the probabilities for the various joint outcomes (i.e., basic joint states).

The outcomes generated by simple product-state superpositions lead to nothing out of the ordinary from a classical perspective, but a different story emerges when one considers the superpositions of entangled states, which are essentially nonseparable, nonfactorizable, multiparticle superposition states. Consider, for example, the following entangled two-photon state for photons A and B:

$$\psi_{AB} = \frac{1}{\sqrt{2}} \{ T_a \otimes T_a \} + \frac{1}{\sqrt{2}} \{ S_a \otimes S_a \}.$$

This expression includes two basic joint states of the $\alpha\alpha$-measurement scenario, each of which has a joint coefficient of $1/\sqrt{2}$ and thus a 50 percent chance of showing up in any $\alpha\alpha$-measurement. As is generally the case with entangled states, the polarization state of each individual photon *cannot* be separately expressed (i.e., the joint state cannot be reexpressed as a product state) under any measurement scenario—the overall state is thus nonseparable or "nonfactorizable." Like all entangled states, this state leads to some peculiar predictions regarding left and right outcomes. For instance, even though both α-outcomes are initially possible for each photon (which, when considered by themselves, are each in a superposition of the two basic α-states), a T_a-outcome on the left "forces" a T_a-outcome on the right and vice versa. Similarly, an S_a-outcome on the left "forces" an S_a-outcome on the right and vice versa. According to quantum theory, influences of this kind are instantaneous and persist no matter how widely separated the entangled particles might become. This is what Einstein called "spooky action at a distance."

Before turning to the quantum prediction for the behavior of photons in a Bell experiment, it will be helpful to remind ourselves briefly of the four key behaviors observed by physicists in actual Bell experiments (all of which were mentioned in Chapter 3).

1. Individual polarization measurements produce apparently random outcomes; neither wing shows any discernable pattern of outcomes over many runs.

2. All four joint outcomes—TT, TS, ST, SS—occur apparently randomly from run to run when the filters are oriented along different but non-orthogonal directions.

3. Joint outcomes tend toward perfect *anti-correlation* as the difference between the left and right filter orientations approaches 0° and toward perfect *correlation* as this difference approaches 90° (i.e., orthogonality), becoming exclusively one or the other in the relevant limit.

4. Joint measurements performed on large collections of jointly produced photons lead to violations of Bell's inequality for some angles.

We are now ready to calculate the quantum prediction for a Bell-type experiment.

5   THE QUANTUM CALCULATION

One of the most important consequences of the quantum approach to state representation in general—and its most notable departure from the classical point of view—is that randomly varying outcomes can be associated with the same initial state. This aspect of the theory is relevant to the quantum treatment of the jointly produced photons used in Bell-type experiments because physicists regard every pair of photons in such an experiment as being in exactly the same initial joint (superposition) state. The word for this, introduced in the previous chapter, is "ensemble." The state of ensembles of photons used in Bell-type experiments, which is known for obscure reasons that I won't go into as the "singlet state,"[12] can be written as follows:

$$\psi_{AB} = \frac{1}{\sqrt{2}} \{ \mathbf{T_v} \otimes \mathbf{S_v} \} - \frac{1}{\sqrt{2}} \{ \mathbf{S_v} \otimes \mathbf{T_v} \}. \qquad 4.10$$

We are now ready to see how ascribing this state to every pair of photons leads not only to predictions that match the cumulative results obtained under various measurement scenarios but also to a violation of Bell's inequality.

We must begin by recalling an important adjustment Bell made to the EPR–Bohm setup. Unlike EPR and Bohm, he allowed each measurement device to assume varying orientations from run to run. So far in this chapter we have confined ourselves to measurement scenarios in which at least one of the filters is oriented along the vertical axis. We must now relax this condition and allow each filter to assume one of two possible orientations: α_1 and α_2 for the left filter, and β_1 and β_2 for the right filter. This aligns the character of our prediction with the inequality developed at the end of Chapter 3. Recall that each of the four orientations can be any angle we choose. In order to keep track of which measurement angle goes with which photon, we will write the angle in Equations 4.7 and 4.8 as "α_1" or "α_2" when referring to the left photon and "β_1," or "β_2," when referring to the right photon. We will omit subscripts entirely during the initial part of the derivation, reintroducing them only when it becomes necessary to uniquely identify the photons' average preferences for a given measurement scenario.

We start by applying the five basic steps for calculating joint probabilities to Equation 4.10. Once we have arrived at the relevant probabilities, we will need to determine how to incorporate them into an expression for the average preference for correlation versus anti-correlation under a given measurement scenario. Recall that the first step involves transforming an initially given joint-state description into an expression relevant to the measurements we actually intend to perform. In our case, this means using Equations 4.7 and 4.8 to reexpress Equation 4.10 in terms of α-outcomes for the left-hand photon and β-outcomes for the right-hand photon. The four relevant substitutions lead to an overall joint state that contains a superposition of two product states, each of which contains two internal superpositions:

$$\psi_{AB} = \frac{1}{\sqrt{2}}\left\{\left(\sin(\alpha)\mathbf{T}_\alpha - \cos(\alpha)\mathbf{S}_\alpha\right) \otimes \left(\cos(\beta)\mathbf{T}_\beta + \sin(\beta)\mathbf{S}_\beta\right)\right\}$$
$$- \frac{1}{\sqrt{2}}\left\{\left(\cos(\alpha)\mathbf{T}_\alpha + \sin(\alpha)\mathbf{S}_\alpha\right) \otimes \left(\sin(\beta)\mathbf{T}_\beta - \cos(\beta)\mathbf{S}_\beta\right)\right\}.$$

Each of these internal superpositions can be eliminated by expanding the expression as before:

$$\psi_{AB} = \frac{1}{\sqrt{2}}\left\{\sin(\alpha)\mathbf{T}_\alpha \otimes \cos(\beta)\mathbf{T}_\beta\right\} + \frac{1}{\sqrt{2}}\left\{\sin(\alpha)\mathbf{T}_\alpha \otimes \sin(\beta)\mathbf{S}_\beta\right\}$$
$$+ \frac{1}{\sqrt{2}}\left\{-\cos(\alpha)\mathbf{S}_\alpha \otimes \cos(\beta)\mathbf{T}_\beta\right\} + \frac{1}{\sqrt{2}}\left\{-\cos(\alpha)\mathbf{S}_\alpha \otimes \sin(\beta)\mathbf{S}_\beta\right\}$$
$$- \frac{1}{\sqrt{2}}\left\{\cos(\alpha)\mathbf{T}_\alpha \otimes \sin(\beta)\mathbf{T}_\beta\right\} - \frac{1}{\sqrt{2}}\left\{\cos(\alpha)\mathbf{T}_\alpha \otimes -\cos(\beta)\mathbf{S}_\beta\right\}$$
$$- \frac{1}{\sqrt{2}}\left\{\sin(\alpha)\mathbf{S}_\alpha \otimes \sin(\beta)\mathbf{T}_\beta\right\} - \frac{1}{\sqrt{2}}\left\{\sin(\alpha)\mathbf{S}_\alpha \otimes -\cos(\beta)\mathbf{S}_\beta\right\}.$$

Next, we contract individual coefficients, being careful to keep track of the various plus and minus signs:

$$\psi_{AB} = \frac{1}{\sqrt{2}}\sin(\alpha)\cos(\beta)\left\{\mathbf{T}_\alpha \otimes \mathbf{T}_\beta\right\} + \frac{1}{\sqrt{2}}\sin(\alpha)\sin(\beta)\left\{\mathbf{T}_\alpha \otimes \mathbf{S}_\beta\right\}$$
$$- \frac{1}{\sqrt{2}}\cos(\alpha)\cos(\beta)\left\{\mathbf{S}_\alpha \otimes \mathbf{T}_\beta\right\} - \frac{1}{\sqrt{2}}\cos(\alpha)\sin(\beta)\left\{\mathbf{S}_\alpha \otimes \mathbf{S}_\beta\right\}$$
$$- \frac{1}{\sqrt{2}}\cos(\alpha)\sin(\beta)\left\{\mathbf{T}_\alpha \otimes \mathbf{T}_\beta\right\} + \frac{1}{\sqrt{2}}\cos(\alpha)\cos(\beta)\left\{\mathbf{T}_\alpha \otimes \mathbf{S}_\beta\right\}$$
$$- \frac{1}{\sqrt{2}}\sin(\alpha)\sin(\beta)\left\{\mathbf{S}_\alpha \otimes \mathbf{T}_\beta\right\} + \frac{1}{\sqrt{2}}\sin(\alpha)\cos(\beta)\left\{\mathbf{S}_\alpha \otimes \mathbf{S}_\beta\right\}.$$

Each basic joint state appears twice in this expression, so we group the coefficients accordingly (we skipped this step in the example given earlier because each basic joint state appeared there only once):

$$\psi_{AB} = \frac{1}{\sqrt{2}}[\sin(\alpha)\cos(\beta) - \cos(\alpha)\sin(\beta)]\{\mathbf{T}_\alpha \otimes \mathbf{T}_\beta\}$$

$$+ \frac{1}{\sqrt{2}}[\cos(\alpha)\cos(\beta) + \sin(\alpha)\sin(\beta)]\{\mathbf{T}_\alpha \otimes \mathbf{S}_\beta\}$$

$$- \frac{1}{\sqrt{2}}[\cos(\alpha)\cos(\beta) + \sin(\alpha)\sin(\beta)]\{\mathbf{S}_\alpha \otimes \mathbf{T}_\beta\}$$

$$+ \frac{1}{\sqrt{2}}[\sin(\alpha)\cos(\beta) - \cos(\alpha)\sin(\beta)]\{\mathbf{S}_\alpha \otimes \mathbf{S}_\beta\}.$$

Here the trigonometric "difference" identities

$$\sin(\alpha - \beta) = \sin(\alpha)\cos(\beta) - \cos(\alpha)\sin(\beta)$$
$$\cos(\alpha - \beta) = \cos(\alpha)\cos(\beta) + \sin(\alpha)\sin(\beta)$$

come to our aid, allowing us to simplify ψ_{AB} as:

$$\psi_{AB} = \frac{1}{\sqrt{2}}\sin(\alpha - \beta)\{\mathbf{T}_\alpha \otimes \mathbf{T}_\beta\}$$

$$+ \frac{1}{\sqrt{2}}\cos(\alpha - \beta)\{\mathbf{T}_\alpha \otimes \mathbf{S}_\beta\}$$

$$- \frac{1}{\sqrt{2}}\cos(\alpha - \beta)\{\mathbf{S}_\alpha \otimes \mathbf{T}_\beta\}$$

$$+ \frac{1}{\sqrt{2}}\sin(\alpha - \beta)\{\mathbf{S}_\alpha \otimes \mathbf{S}_\beta\}.$$

4.11

As unlikely as it may seem, Equations 4.10 and 4.11 represent exactly the same joint state—the singlet state—the only difference being that Equation 4.10 expresses this state specifically in terms of vv-measurement, whereas Equation 4.11 expresses it more generally in terms of any α-measurement on the left and any β-measurement on the right. Squaring each joint coefficient leads, as always, to the quantum mechanical probability of each joint outcome:

$$P_{QM}\left(\mathbf{T}_\alpha \otimes \mathbf{T}_\beta\right) = \frac{1}{2}\sin^2(\alpha - \beta)$$

$$P_{QM}\left(\mathbf{T}_\alpha \otimes \mathbf{S}_\beta\right) = \frac{1}{2}\cos^2(\alpha - \beta)$$

$$P_{QM}\left(\mathbf{S}_\alpha \otimes \mathbf{T}_\beta\right) = \frac{1}{2}\cos^2(\alpha - \beta)$$

$$P_{QM}\left(\mathbf{S}_\alpha \otimes \mathbf{S}_\beta\right) = \frac{1}{2}\sin^2(\alpha - \beta).$$

4.12

The first thing to notice about these probabilities is that they include each of the basic polarization vectors for each individual photon: the T_α and S_α vectors for an α-measurement on the left photon, and the T_β and S_β vectors for a β-measurement on the right photon. This means that, in general, each individual outcome has a nonvanishing probability of appearing in any joint measurement. What may not be immediately apparent is that each individual outcome will also appear unpredictably from run to run. That is to say, the probability we get for each individual outcome, ignoring what happens in the distant wing (the so-called "marginal" probability for each wing), is 1/2. We can see that this is true for the individual outcome S_α in the left wing by adding together the probability of each joint state containing this outcome (the third and fourth terms):

$$P_{QM}(S_\alpha) = \frac{1}{2}\cos^2(\alpha - \beta) + \frac{1}{2}\sin^2(\alpha - \beta) = \frac{1}{2}\left[\cos^2(\alpha - \beta) + \sin^2(\alpha - \beta)\right] = \frac{1}{2}.$$

Similar calculations confirm the same result for the other three individual states. Equation 4.11 thus successfully predicts the first observed behavior of jointly produced photon pairs: Each *individual* photon behaves unpredictably when considered by itself.

Now imagine that the orientations of the two filters are neither perfectly parallel nor perfectly orthogonal. In this case, none of the four basic joint outcomes disappears entirely, in agreement with the second observed behavior: All *joint* outcomes occur randomly from run to run when the filters are oriented along different but nonorthogonal directions. On the other hand, as we reduce the difference between β and α to $0°$, the coefficients of the second and third terms converge to $1/\sqrt{2}$ and the first and fourth terms go to zero (see the values for sine and cosine in Figure 4.3), leaving only anti-correlated outcomes. And, if we increase the difference to $90°$, the coefficients of the first and fourth terms converge to $1/\sqrt{2}$ and the second and third terms go to zero, leaving only correlated outcomes. Equation 4.11 thus successfully predicts the third key behavior of photon pairs as well: Joint measurements tend toward anti-correlations as the difference between filter orientations approaches $0°$ and toward correlations as the difference approaches $90°$.

We are now ready to compare the quantum prediction with the central mathematical result of Chapter 3, i.e., the classically motivated Bell-type inequality given in Equation 3.18:

$$\left| \langle A(\alpha_1)B(\beta_1) \rangle + \langle A(\alpha_1)B(\beta_2) \rangle + \langle A(\alpha_2)B(\beta_1) \rangle - \langle A(\alpha_2)B(\beta_2) \rangle \right| \le 2.$$

This inequality brought together into a single expression the source's average preference for correlation versus anti-correlation under all four measurement scenarios. We now want to generate a theoretical prediction for the average associated with each of the four possible measurement scenarios, using the quantum probabilities obtained in Equation 4.12. In this case, we can no longer speak unproblematically about the "source's preference" because, according to the standard interpretation, the source is determining only the overall joint-polarization state of each photon pair, not the individual polarization state of either member of the pair. Thus we should speak instead of the average preference for correlation versus anti-correlation displayed by a collection of photon pairs under a particular measurement scenario. The relationship between this average preference and the probability of each joint outcome under this scenario is not a straightforward one, but neither is it overly complicated. In order to see the connection clearly, we must briefly acquaint ourselves with the idea of a "weighted sum."

Consider the following situation. Every day during the summer Amanda walks to and from the store to buy fresh bread. On rainy days, she takes a one-block route. When it doesn't rain, she takes a more scenic eight-block route. As it happens, Amanda lives in a part of the world where it rains, on average, five out of every seven days during the summer. This means that the likelihood of rain on a given day is 5/7 (71 percent), whereas the likelihood of no rain on a given day is 2/7 (29 percent). If we multiply or "weight" the length of each route by the likelihood that it will be taken, we can predict the average distance Amanda will walk each day as a sum of the two "weighted" trip lengths:

$$\left(2\frac{\text{blocks}}{\text{day}} \times \frac{5}{7}\right) + \left(16\frac{\text{blocks}}{\text{day}} \times \frac{2}{7}\right) = \text{average of } 6\frac{\text{blocks}}{\text{day}}.$$

A weighted sum is a handy tool for predicting averages when the likelihood of each possible outcome is known in advance. We can use this tool, along with the probabilities derived earlier, to generate a quantum theoretical prediction for the value of Equation 3.18.

Recall that we need to calculate not one but four averages, each of which corresponds to the average preference for correlations versus anti-correlations displayed by the sub-ensemble of photon pairs measured under a particular measurement scenario. We thus need to construct a weighted sum for each measurement scenario, first, by replacing the lengths of Amanda's routes with the numerical values associated with the various individual and joint-polarization outcomes (recall that we assigned +1 and –1 to "Transmit" and "Stop," respectively) and then by replacing the likelihood of Amanda taking each route with the quantum prediction for the likelihood of finding each joint outcome under a particular measurement scenario. Combining these two steps (and reintroducing subscripts to reconnect the average to a specific measurement scenario), we find that the quantum mechanical prediction for the photon pairs' average preference under scenario $\alpha_1\beta_1$ is

$$\langle AB(\alpha_1\beta_1)\rangle_{QM} = (+1)(+1)\times P_{QM}\left\{\mathbf{T}_{\alpha_1}\otimes\mathbf{T}_{\beta_1}\right\}$$
$$+(+1)(-1)\times P_{QM}\left\{\mathbf{T}_{\alpha_1}\otimes\mathbf{S}_{\beta_1}\right\}$$
$$+(-1)(+1)\times P_{QM}\left\{\mathbf{S}_{\alpha_1}\otimes\mathbf{T}_{\beta_1}\right\}$$
$$+(-1)(-1)\times P_{QM}\left\{\mathbf{S}_{\alpha_1}\otimes\mathbf{S}_{\beta_1}\right\}$$

When we substitute the probabilities from Equation 4.12 and simplify terms, this reduces to

$$\langle AB(\alpha_1\beta_1)\rangle_{QM} = \sin^2(\alpha_1 - \beta_1) - \cos^2(\alpha_1 - \beta_1).$$

We can use the trigonometric "double angle" identity

$$\cos(2\theta) = \cos^2(\theta) - \sin^2(\theta)$$

(equating θ with $\alpha - \beta$), to write this expression even more compactly:

$$\langle AB(\alpha_1\beta_1)\rangle_{QM} = -\cos(2\alpha_1 - 2\beta_1). \qquad 4.13$$

Notice that this prediction is not in the form of a probability, even though we used probabilities to construct it. Instead, this prediction is in the form of an average—the average of joint outcomes or "average preference for correlation versus anti-correlation" under measurement scenario $\alpha_1\beta_1$, which is what we considered from an experimental point of view in Chapter 3. We could compare the value of this average directly to the one obtained through repeated runs of the experiment under scenario $\alpha_1\beta_1$, but this wouldn't tell us much. What we really want is to write down the quantum version of Equation 3.18 so that we can see whether the quantum prediction abides by our Bell inequality. Then we can compare the quantum prediction with the results of actual experiments. Equation 3.18 contains four distinct averages, so we need to replace each term with its own quantum version of Equation 4.13 (corresponding to one of the four measurement scenarios: $\alpha_1\beta_1$, $\alpha_1\beta_2$, $\alpha_2\beta_1$, and $\alpha_2\beta_2$). Thus,

$$\left|-\cos(2\alpha_1 - 2\beta_1) - \cos(2\alpha_1 - 2\beta_2) - \cos(2\alpha_2 - 2\beta_1) + \cos(2\alpha_2 - 2\beta_2)\right| \leq 2. \quad 4.14$$

The left-hand side of this inequality is the quantum mechanical prediction we have been aiming at. How does it behave? In some cases it obeys the inequality. For example, if we let $\alpha_1 = -\beta_1 = 10°$ and $\alpha_2 = -\beta_2 = 30°$, the left-hand side becomes

$$\left|-\cos(40°) - \cos(80°) - \cos(80°) + \cos(120°)\right| = \left|-0.77 - 0.17 - 0.17 - 0.50\right| = 1.61.$$

However, if we adjust the first pair of angles slightly, so that $\alpha_1 = -\beta_1 = 5°$ and $\alpha_2 = -\beta_2 = 30°$, then we get

$$\left|-\cos(20°)-\cos(70°)-\cos(70°)+\cos(120°)\right| = \left|-0.94-0.34-0.34-0.50)\right| = 2.12,$$

in violation of the classical limit given by Equation 3.18. The significance of this small violation is enormous. Even such a minor difference between the quantum prediction and the classical limit might have meant the end of quantum theory, at least in its present form. But this is not what happened. As we saw at the beginning of the chapter, experiments with real photons and other quantum particles have produced results in excellent agreement with the quantum prediction. Apparently, the trouble lies not with quantum theory, but instead with some combination of the ideas that motivated our Bell-type inequality: property definiteness, state separability, and cause locality.

Which of these three ideas is most seriously threatened by the experimental violations of Bell-type inequalities? The fact that a measurement of one particle appears to force an immediate change in another distant particle in the case of entangled states seems to imply that cause locality may be at risk—recall Einstein's concern about spooky action at a distance. There are at least two reasons, however, to think that entanglement is not best interpreted as a "nonlocal" phenomenon. The first is empirical: Despite the confirmation of quantum entanglement in the correlated behavior of quantum particles, no particle or causal process has ever been observed, directly or indirectly, to move at speeds exceeding that of light in free space. On the contrary, the light limit has been confirmed many times over and is now routinely factored into technologies operating across the globe—from the mundane reality of wireless communications to the more rarefied realm of particle accelerators. The second reason is theoretical: Following Bell's initial discovery, a variety of mathematical proofs have been developed indicating that quantum correlations cannot, even in principle, be used to send superluminal signals (for an example of a "no-signaling theorem," see Cushing 1994, chap. 10, appendix 2; cf. Scarani 2006, 65).

The disagreement between experimental results and our Bell-type inequality must have to do with one or both of the remaining assumptions—property definiteness and state separability. As we saw in the previous chapter, the standard interpretation of quantum theory is highly permissive when it comes to different types of states. Not only does it embrace *property indefiniteness* by ascribing superpositions to individual systems, it also embraces *nonseparability* by refusing to restrict joint states to product states. In other words, the standard view of quantum theory rejects *both* of the remaining classical assumptions. As one might suspect, the peculiar nature of nonseparable quantum states has attracted the attention of a number of philosophers of physics. Abner Shimony calls the tag-team effort on the part of scientists and philosophers to understand the meaning of such states "experimental metaphysics" (1989, 26–27). His point is not to suggest that metaphysical disputes can be settled by experimental outcomes, but rather that scientific input may be helpful and even necessary for gaining a better understanding of reality. In the next chapter we examine several philosophical interpretations that have been developed with regard to quantum theory in general, and quantum entanglement in particular, in order to see both what the hermeneutical task requires and how much interpretive disagreement actually exists.

5 Philosophical Perspectives

A new epoch in physical science was inaugurated . . . by Planck's discovery of the *elementary quantum of action*, which revealed a feature of *wholeness* inherent in atomic processes, going far beyond the ancient idea of the limited divisibility of matter.

—Niels Bohr (1987, 2; italics in original)

Quantum wholeness . . . is a fundamentally new kind of togetherness, undiminished by spatial and temporal separation. No causal hookup, this new quantum thing, but a true mingling of distant beings that reaches across the galaxy as forcefully as it reaches across the garden.

—Nick Herbert (1985, 19)

As Popper has remarked, our theories are "nets designed by us to catch the world." We had better face up to the fact that quantum mechanics has landed some pretty queer fish.

—Michael Redhead (1987, 169)

David Mermin (1985, 41) once noted that physicists who are not deeply bothered by the empirical violation of Bell's inequality either miss the point entirely, make assertions that can be shown to be false, or glibly claim that Niels Bohr straightened the whole thing out long ago. People *should* be bothered, thinks Mermin, and he is not alone. Henry Stapp (1977, 173; cf. Whitaker 1998) contends that the confirmation of the quantum prediction for Bell-type experiments constitutes one of the most profound scientific discoveries ever made. If Mermin and Stapp are right, then Bell-violating correlated behavior is no small cloud on the experimental horizon but a storm front that portends a drastic change in how we understand the physical world. The weather, however, is unsettled. Although quantum theory accurately accounts for all of the correlated behaviors thus far observed in Bell-type experiments, neither this nor any other scientific theory can dictate the meaning of its own success. The mathematics of quantum theory that has led to the idea of entanglement provides neither a deeper explanation of the correlations nor a clearer account of their metaphysical significance (Howard 1989, 228). To investigate their meaning demands that we add yet another layer of understanding to those we have obtained thus far. We must now venture beyond the exacting realm of scientific experimentation, beyond the abductive realm of scientific theorizing, and into the speculative realm of philosophical interpretation. We must join those physicists and philosophers who, unwilling to content themselves with predictive success, have attempted to say something about the ontological implications of the series of discoveries that has led physicists to use the term "entanglement" when referring to the correlations they observe in Bell-type experiments. This is controverted territory, but it is unavoidable for anyone wishing to go beyond the confines of bare empiricism.

Whereas Chapter 3 focused on three classical ideas that are jointly contradicted by the Bell-violating data amassed since the 1970s, the present chapter explores the interpretive possibilities opened up by the predictive achievements of quantum theory with regard to the very same data. A significant body of philosophical literature is accumulating around the meaning of entanglement, some of it from a relational-holist perspective. An important task of this chapter is to catalogue the variety of ways in which people have begun to think about the matter so that, with the broader range of options before us, we can situate the relational-holist view vis-à-vis the larger conversation and understand its particularities more clearly (my own account of the options is informed by Jones 2002). I do not want to argue for the absolute superiority of the relational-holist interpretation over the others. To do so would be disingenuous, given the nascent state of the literature. None of these interpretations has yet been developed to the point where one can rightfully be selected as being obviously superior to its competitors. I want to argue instead for a more modest pair of claims, namely, that the relational-holist view is at least as plausible as the others and that each of the other views either has difficulties with self-consistency or comes with a more burdensome metaphysical price tag.

Although each of the views discussed in the following is indeed highly speculative, seen together they look less like arbitrary flights of fancy and more like pioneering responses to what is a moment of considerable, perhaps even Copernican, significance in the history of science. It will be helpful in this regard to begin by reviewing the larger debate over the meaning of quantum theory itself. The first section thus picks up where the previous chapter left off with regard to the central interpretive problem associated with quantum theory, i.e., the measurement problem. The next four sections present and evaluate several competing interpretations of entanglement, concluding with the relational-holist interpretation. The final section of the chapter anticipates the theological discussion of the next chapter by examining several of entanglement's most relevant characteristics from a relational-holist point of view.

In the decades since Bell published his inequality, physicists and philosophers have scrutinized every aspect of the underlying argument. There is little doubt at this point as to the soundness of Bell's conclusion. In fact, his original argument has only been strengthened and generalized in subsequent work (see, for example, Greenberger et al. 1990; Hardy 1993; Cabello 1998;

for an especially clear account of the implications deriving from the work of Greenberger et al., see Bub 2010, 21–24); after much debate it would still appear to be the case that the classical worldview cannot easily be reconciled with the predictions of quantum theory (but see Fine 1996; Winsberg and Fine 2003). Both the academic and popular literature on Bell's work have grown exponentially since the mid-1980s.[1] However, despite the fact that entanglement is now a well-understood phenomenon from a mathematical and experimental point of view, its ontological implications regarding the nature of physical processes are far from obvious. Early interpreters followed Bell in labeling quantum correlations "nonlocal," but as we saw in Chapter 3 the term "nonlocality" is usually construed in terms of superluminality, especially the possibility of superluminal signaling. As quantum theory does not appear to make superluminal signaling possible, the term "nonlocality" has largely fallen out of favor as the descriptor of choice. As we have seen, the more recently preferred term, "entanglement" [*Verschränkung*], was coined by Erwin Schrödinger in the early decades of the twentieth century (1935). Ask a physicist on the street today why quantum particles violate Bell's inequality and he or she is likely to respond matter-of-factly, "Because they are entangled." But what does it mean to be entangled? How should we understand the correlated behavior of photons and other quantum particles in Bell-type experiments? We must begin our attempt to answer these questions by revisiting and then extending the previous chapter's discussion of the so-called "measurement problem."

1　WHOSE THEORY, WHICH INTERPRETATION?

In Chapter 3 I spoke matter-of-factly about the "standard interpretation" of quantum theory. I defined this view as one that broadly affirms the realist intent of scientific theorizing in general, the ontologically indeterministic and indefinite character of at least some physical processes in particular, and the change or "collapse" of one or more of a system's indefinite properties into definiteness upon some sort of interaction between the system and its environment (typified by the act of measurement). Although this view is commonly attributed to Niels Bohr and frequently referred to as the "Copenhagen interpretation" in honor of the Dane, it is more correctly linked to the views of Werner Heisenberg—one of Bohr's students—who thought about quantum mechanical processes in explicitly Aristotelian terms, i.e., of quantum systems and their properties as moving from potentiality to actuality (see, for example, 1958). I will continue to use the phrase "standard interpretation" in the sense defined earlier because this basic viewpoint continues to be the default interpretive position of many physicists (and is frequently employed as a foil in arguments for alternative perspectives). The time has come, however, to give some indication of the complexity of the interpretive debate.

Recall that the standard interpretation offers no physical account of how objects transition from indefinite to definite states. In fact, one could argue that the mathematical formalism leaves no room for such transitions. Squaring this point with the fact that, in everyday life and in the laboratory, we always find physical objects in apparently definite states, is the heart of the measurement problem (see Butterfield 2001). This conceptual difficulty has generated a large number of solutions. The fact that they have a commonly radical character testifies to the seriousness of the matter. In this chapter I focus only on representative proposals within three of the main interpretive categories: realism within the existing formalism, realism with a modified formalism, and anti-realism (for a broader survey, see Albert 1992; d'Espagnat 2006). Most physicists and philosophers who feel compelled to wrestle with the measurement problem are motivated by the realist conviction that scientific theories are successful because they model, to some degree of accuracy or another, the actual way the world works and is put together. In what follows, I give special attention to proposals from the realist perspective while noting the basic rationale of the anti-realist position.

To begin, it is worth noting that no realist interpretation of quantum theory can avoid the interpretive burden posed by the empirical confirmation of the quantum prediction for Bell-type experiments. This turn of events is certainly a conundrum for any realism as strong as Albert Einstein's. If the experimental results at the two ends of a Bell apparatus jointly imply that the measurements cannot be the straightforward disclosures of the particles' previously existing and definite states, well, what then do they imply? It is also a conundrum for Heisenberg's neo-Aristotelian view, in which the dynamical properties of quantum objects evolve from fuzzy potentialities to definite actualities. How, after all, does the process of actualization in one place superluminally trigger the actualization of properties in distant places? Finally, the empirical confirmation of quantum theory in the case of Bell-type experiments is even a conundrum for the difficult-to-grasp realism of Niels Bohr. Although Bohr made various pronouncements on the subject, not all of which are easily reconcilable with each other, he appears at the very least to have thought that an object's physical properties can be defined only in accord with the character of the experimental context (Redhead 1987, chap. 4; for a discussion of Bohr's realism, see Folse 1989). But then how can a change in a nearby experimental context immediately affect what is well defined about a quantum object in some remote location? Notwithstanding the differences between each of these views, they all share a basic realist commitment that will not allow them to sidestep the question of how to understand the results of Bell-type experiments.

Of course it is possible to reject the realist perspective. Then there would be no reason to suppose that quantum theory's peculiar use of the principle of superposition has any bearing one way or the other on the manner in which physical objects carry their properties (Norris 2000). I remain a realist because I, like others, regard the cost of adopting an anti-realist stance vis-à-vis science as unreasonable, i.e., without the realist presumption the predictive power of successful scientific theories becomes a mystery (Putnam 1975, 73; Musgrave 2006–2007). There is more than a kernel of sense, however, in the anti-realist arguments (e.g., van Fraassen 1980; Cartwright 1983); even a realist can admit that there is more to reality than we catch in our empirical nets. But to deny the possibility of drawing ontological inferences from scientific theories on the grounds that they cannot be expected to bear the burden of guiding our interpretations of the world is to undercut the central motivation animating most scientists' pursuit of knowledge: to know the world better. When anti-realist caution is elevated to the level of a philosophical program, it gets an undeserving free ride on the backs of the vast majority of scientists whose realist convictions push science forward. This is not to say, however, that those who call upon the "truth" of any particular theory can ignore the anti-realist's warnings. The tension between the realist and the anti-realist positions has proven crucial to the production of reliable knowledge. The epistemological relationship between these two voices parallels the semantic simultaneity of the "is" and "is not" of metaphor, fostering a tension that keeps us epistemologically humble even as it drives us to further exploration. Like all scientific theories, quantum mechanics is grist for the mill as we seek to assemble truth-naming, meaning-sustaining interpretations of the world. The fact that theories and interpretations are inevitably incomplete and susceptible to revision, even replacement, does not preclude a critical but nonetheless realistic attitude toward theories we judge to be the best at describing some portion of reality. The same point has been cogently argued with regard to theological theories, in which case the meaning of "best" must be carefully tuned to the particular nuances of the experiential realm over and against those of the experimental realm (Tillich 1951, 102; for a discussion and defense of chastened, i.e., "critical," realism in science as well as theology, see Barbour 1971, 1997; Peacocke 1984; Van Huyssteen 1989; Polkinghorne 2005).

Realist interpreters of quantum theory have in general pursued one of two basic strategies for solving the measurement problem: (1) develop a new ontological interpretation of the theory's existing mathematical formalism, or (2) modify the theory's existing formalism in accord with some principle that has *prima facie* plausibility (cf. Stapp 1989; McMullin 1989, 300–302). Advocates of various proposals within each of these two strategies have been debating one another for almost a century now, and a clear winner has yet to emerge. My aim in this initial section is not to settle the debate but to alert readers to its ongoing existence and to anticipate its relevance for the coming exploration of entanglement as a theological metaphor.

Perhaps the most widely discussed realist alternative that operates wholly within the confines of the received mathematical formalism is the family of interpretations built upon the early ideas of Louis de Broglie. Expanding de Broglie's initial work, David Bohm (1952) developed a definite-value interpretation of the quantum formalism by recasting the meaning of the basic mathematical object of the theory, the "wavefunction."[2] He argued that the wavefunction guides the particle *through* space—de Broglie had called it a "pilot wave"—rather than describing the particle's state *in* space. The ontology in this case expands from a quantum particle in a superposition described by a wavefunction to a classical particle in a definite state plus its guiding pilot wave (the wavefunction ontologized). The experimental setup dictates the character of the wavefunction, which in turn dictates the character of what Bohm called "the quantum potential." He thought of the latter as a cosmically ubiquitous field capable of *nonlocally* influencing all objects in the universe by connecting them to one another instantaneously.[3] If one changes the experimental setup, then the quantum potential changes accordingly and instantaneously across the entire universe. In this way the de Broglie–Bohm interpretation allows all physical objects to have definitely valued properties. It does so, however, at the cost of linking the motion of one particle nonlocally via the quantum potential to the configuration of all the other particles in the universe at the same instant. Because Bohm's view follows from a rearrangement of the math contained in the standard formalism, it generates exactly the same predictions as the standard formalism, including a violation of Bell's inequality. At the interpretive level, however, the de Broglie–Bohm approach trades indefiniteness for nonlocality (which resides in the "entangling" character of the quantum potential; see Cushing 1994, 62) inasmuch as a change in the state of one particle instantly (i.e., superluminally) affects the state of every other particle in the universe.

One might be surprised to learn that superluminal signaling does not automatically follow from Bohm's rearranged mathematics. Although cause locality is violated both in principle and in fact in Bohm's universe (for a dissenting view, see Dickson 1998), a practical barrier to superluminality still exists within Bohm's view, namely, our ignorance of the precise state of any object at any given time. Such ignorance is unavoidable from the Bohmian perspective, and it entails the now familiar statistical character of quantum predictions. In the Bohmian case, however, the indeterminism (as well as the indefiniteness) of physical processes is interpreted epistemologically rather than ontologically. Accordingly, Bohm's universe is both definite and determined, in addition to being nonseparably entangled and uncontrollably nonlocal through the quantum potential.

Recall that it was Bohm's discovery of a nonlocal, property-definite version of quantum theory that prompted Bell to think about whether any *local* property-definite theory could reproduce all of the predictions of quantum mechanics. Bell's own work confirmed that Bohm's theory was not an exception but rather the rule. Any property-definite theory capable of reproducing the statistics of quantum theory would have to be entangled (in Bell's language, "nonlocal"). Bohm's approach has been widely criticized as metaphysically extravagant and at odds—at least in principle if not in practice—with special relativity,[4] but Bohm nonetheless deserves credit for teasing out the implications of de Broglie's early attempt to save property definiteness within the quantum framework (for a criticism of Bohm's detractors, see Cushing 1994) . Tim Maudlin (2002, 121; italics in original) states the matter forcefully (setting terminological quibbles aside) when he says: "Violations of Bell's inequality show that the *world* is non-local. It can be no criticism of a theory that it displays this feature of the world in an obvious way." The fact that entanglement is present in its own way within the standard interpretation only reinforces Maudlin's point. In subsequent sections I will not give significant attention to the particularities of Bohm's highly developed holist perspective on quantum theory (for helpful discussions, see Russell 1985; Sharpe 1993), not because I find his thoughts on this subject unappealing or because Bohm's holism is more radical than the relational holism I consider later, but because the problem of what entanglement means is not unique to Bohm's approach. As I mentioned earlier, *all* realist views—whether deterministic or indeterministic—must face the question of how to understand the experimental violation of Bell's inequality.

John Cramer (1986, 647–687) has developed another realist alternative to the standard interpretation that operates within the confines of the received mathematical formalism. His proposal, which he calls the "transactional interpretation," attempts to circumvent the holist implications of entanglement by appealing to the notion of *backward causation*. At the heart of Cramer's proposal is the idea that a detector (a measurement device) in the future interacts *from the future* with a source in the present by sending information about itself backward in time to the source. The source (in the present) and the detector (in the future) cooperatively determine the probability of a given outcome through their interaction, which Cramer calls "the transaction." The alleged benefit of this approach is that the components of the experimental apparatus need only communicate with one another at or below the speed of light. The transactional interpretation thereby avoids both superluminal signaling and nonclassical, holistic relationships between particles.

One interesting feature of Cramer's interpretation is that the interaction between source and detector happens, whether deterministically or otherwise, not at a moment in time but at many moments (all together?) across time. The transaction can be understood as a "trans-temporal" or perhaps "a-temporal" process. The relational-holist approach to be discussed in the following avoids backward-in-time signaling through its use of state non-separability, but it too suggests a trans-temporal connection, as we shall see. Cramer's appeal to backward causation is a good example of the length to which interpreters will go to solve the various conundrums associated with the standard interpretation of quantum theory.

According to the so-called "many-worlds interpretation," initiated by Hugh Everett and later developed by Bryce DeWitt (DeWitt, Graham, and

Everett 1973), the universe itself can be described by a wavefunction. Each possibility contained within the universe's wavefunction is realized as a newly born universe or a new branch of the previously existing universe as it continually splits upon the occasion of each quantum event. One variant of this view (Grib and Rodrigues 1999, chap. 6; Bell 1982, 611–637) posits an infinite number of pasts as well as an infinite number of futures to avoid violating the conservation of mass–energy, which is a problem for the forward-only account because of the constant production of new universes. In the many-worlds interpretation an observer experiences a collapsing wavefunction not because it actually collapses but because of the limited nature of consciousness, which forces an individual observer to identify one of infinitely many presents as *the* present (and, possibly, one of infinitely many pasts as *the* past).

Another variant of the many-worlds interpretation is the so-called "many-minds" approach. Here the universe's wavefunction always describes an indefinite state, always points to a superposition of realities. The wavefunction *never* collapses and measurement only *appears* to produce definite results (Albert and Löwer 1988, 195–213; 1989, 121–138; Albert 1992; cf. the essays by Berry and Butterfield in Russell et al. 2001). It is the special status of consciousness that allows the mind to *perceive* only one definite state of affairs. Because there are no definite outcomes in this view, there are no correlations to be accounted for and thus no entanglement. However, an *infinity* of minds must now be posited for each body in order to make the approach consistent with the fact that infinitely many outcomes are typically possible in any given quantum event. The many-worlds and many-minds approaches differ in detail, but both manifest what John Polkinghorne (2002, 53) has called an "astonishing ontological prodigality" in their willingness to countenance the proliferation of universes or minds. According to these scenarios, although the stories told by observers will inevitably include "strange" phenomena such as entanglement, Bell's inequality is *not* really violated by these phenomena because all possible outcomes have occurred in the past and will occur in the future—whether in many universes or in many minds. As with the transactional interpretation, this family of interpretations ascribes the seemingly unexplainable quantum "correlations" witnessed in the laboratory to the apparently limited scope of consciousness (see Maudlin 2002, 5, 218–220; cf. d'Espagnat 2006, 239 for an explicitly neo-Kantian appeal to the limits of perception).

Rather than attempt to solve the measurement problem by reinterpreting quantum theory's mathematical formalism, some have sought to modify the received formalism by replacing Schrödinger's linear equation with a nonlinear one (e.g., Shimony 1989) or by introducing an actual mechanism for collapse (e.g., see Ghirardi, Rimini, and Weber 1986, 470–491; Penrose 2005, chaps. 29, 30). In the latter, the so-called "dynamical state-reduction" models, the rate of collapse is typically linked to the amount of matter present. As the amount of matter increases, so does the likelihood that the state of the system will collapse onto some definite collection of values. Thus, the first measurement at one end of a Bell-type experiment merely entangles both particles with the measurement apparatus, whereas the subsequent increase in the amount of matter in the entangled state induces the collapse of the overall wavefunction. After collapse, the now-definite properties of the distant particle lead to an appropriately correlated result in the distant wing. The additional mathematics in this approach turns what was an axiomatic presumption in the standard interpretation—the mysterious collapse of the wavefunction—into a concrete physical mechanism. Having such a mechanism for collapse solidifies the place of entanglement within quantum theory, but it does little to render the phenomenon of correlated behavior more comprehensible; collapse is still an instantaneous and global event that connects distant objects in ways that appear to breach the relativistic structure of space-time (Dickson 1998, 182–183; cf. Grib and Rodrigues 1999, 147).

To summarize, by the middle of the twentieth century, the indeterministic collapse of the wavefunction had come to be regarded as the signature feature of quantum theory. Bohm's discovery of a deterministically interpretable version of the mathematical formalism cast doubt on the inevitability of the standard interpretation, but it was for the most part not taken seriously as an interpretive competitor. Each of the alternatives discussed earlier has been proposed specifically as a solution to the measurement problem, and as we saw each alternative also offers its own view of entanglement. As the idea of entanglement has increasingly supplanted the idea of indeterminism as the defining feature of quantum theory, the one that sets it apart from all classical theories, the broader interpretive debate over the meaning of quantum theory has increasingly become a debate over the specifics of entanglement. We now take up three specific proposals for interpreting its physical significance. Each comes from an analysis of the idea of entanglement rather than a broader interpretation of quantum theory, although any of the three might be incorporated into a more general view.

2   DO WE NEED TO UNDERSTAND?

Some physicists and philosophers have cautioned against reading too much into quantum correlations. David Mermin, despite his sense that entanglement is not something to be taken lightly, has warned against leaping from "the inability of physics to deal with the individual properties of an individual system, to the miraculous creation of such properties from afar" (1999, 572). Mermin rejects Abner Shimony's phrase, "passion-at-a-distance" (Shimony 2001, 10, 11), a slogan for a weaker kind of connection between physical objects than everyday causal connection. The path forward, according to Mermin, lies in the direction of a chastened metaphysics of particle properties, one that views "physical reality as comprised only of correlations between different parts of the physical world, and not at all of unconditional properties possessed by those parts" (1999, 582; cf. Hughes 1989a). This reluctant metaphysician suggests instead the term "fashion-at-a-distance," but even he concedes that metaphysical adjustments may be required.

Arthur Fine has leveled a more serious objection against attempts at "understanding" entanglement. As Fine sees it, "the 'natural' or zero state of mind with regard to the correlations between outcomes in different wings of [a Bell-type] experiment should be no different from what it is for the randomness of the outcomes in each wing separately; namely, unless features are present that point to the correlations (but not the randomness) as in need of explanation, they are not" (1989b, 181). Fine thinks that it is inconsistent to allow, on the one hand, for "objective randomness" in the case of individual quantum events (a key aspect of the standard interpretation) and yet to demand, on the other, an explanation of correlations between events. Both individual and correlated behaviors, he points out, are embedded within the quantum formalism in such a way as not to have a cause. Fine asks, "Why, from an indeterminist perspective, should the fact that there is a pattern *between* random sequences require any more explaining than the fact that there is a pattern internal to the sequences themselves?" (1989b, 192; italics in original). But the answer is not hard to find. If one knew that two seemingly random sequences of data had been generated by a common cause, one would not be surprised to find some kind of correlation between them. But if there were good reason to believe that no common-cause explanation existed, e.g., if they were produced at "space-like" separation from one another (see Chapter 3, note 11, this volume), then their nonrandom correlation would indeed be something to puzzle over—especially if one had good reason to believe that each sequence had been generated by some ontologically indeterministic process. Fine is right to insist that we should "see patterns *between* sequences as part of the same natural order as patterns *internal* to the sequences themselves" (1989b, 193; italics in original), but nonrandom patterns between apparently random sequences demand special consideration when by other accounts the events that produced the two sequences cannot be causally connected (Teller 1989, 222; Zeilinger 2010, 36).

Perhaps the past empirical successes of theories that provide us with explicitly causal accounts of physical phenomena have led us to unjustified expectations. Past success is no guarantee, after all, that causal explanations will work in all cases. "There does not have to be a reason for everything," says Bas van Fraassen, who is content to let quantum correlations go unexplained (1989, 113, 109). Quantum theory, van Fraassen thinks, may just be an example of a scientific theory that achieves empirical adequacy *without* going the route of causal explanation. James Cushing put the same point more sharply: "Are explanation and understanding really possible when a detailed causal explanation is in principle impossible?" (1989, 14). The truth in the rhetoric of Fine, Van Fraassen, and Cushing is that quantum correlations in no way force us to adopt any particular view of reality (or, for that matter, of causality).[5] The present work, however, is motivated by the idea that even the seemingly arcane peculiarities of quantum entanglement invite us to deepen our critical-realist understanding of reality.

3   ENTANGLEMENT AT A DISCOUNT

Instead of eliminating the need for understanding, perhaps we ought to search for a physical mechanism that could account for entangled behavior through some minimal departure from the classical worldview or the standard interpretation of quantum theory. In this regard, let us return to the principle of cause locality, introduced in Chapter 3. Although commonly interpreted as a prohibition against superluminal signaling, the principle is more correctly understood as a limit on the speed of all *subluminal* physical processes. In fact, Einstein's special theory of relativity allows for physical objects or processes that move or unfold at speeds faster than that of light (see Dickson 1998; Maudlin 2002). What Einstein's theory prohibits is the deceleration of such objects and processes to or below c—just as it prohibits the acceleration of anything moving slower than c from accelerating to or above c. The speed of light is not an absolute limit but an absolute divider. Scientists have dubbed superluminality the regime of "tachyons" (from the Greek ταχύς, meaning "fast"). Although positing such hypothetical particles sounds like a promising way to begin a nearly classical account of quantum correlations, it is important to note that tachyons, if they exist, cannot be very fast but otherwise ordinary particles. Tachyons must have a host of odd properties in order to fit within the relativistic framework. For example, one would need to add energy to a tachyon to slow it down, and the temporal dimension of a tachyon's world is related to the spatial dimensions of our own (and vice versa; for details, see Maudlin 2002). It is not clear, then, how much of the classical worldview could actually be "saved" by using tachyon exchange to understand quantum correlations, but let's set aside such qualms for the moment in order to consider whether tachyons might at least offer a logically coherent basis for understanding entanglement.

How would a tachyon mediate the quantum correlations in a Bell-type experiment? As a superluminal particle, it would be able to carry information about the measurement apparatus and/or the result in one wing superluminally to the other wing. One could then account for correlated behavior between space-like separated events without needing to invoke ideas like "action at a distance" or "relational holism." The difficulty with this approach is that the viability of its candidacy stands or falls with the injunction that observers in different reference frames must never tell inconsistent stories about where a tachyon gets and delivers its information. The problem here is that one cannot guarantee narrative consistency among all of the possible reports of superluminal communication in a universe where special relativity provides a robust (i.e., complete) description of the structure of space and time. In particular, someone in one reference frame will rightfully conclude that the tachyon carries information from wing B to wing A, even though someone else in another reference frame will rightfully conclude just the opposite, i.e., that the same tachyon carries information from wing A to wing B. Mutually inconsistent stories about where the tachyon gets and delivers its information are inevitable.[6] This conceptual issue applies to superluminality in general, but in the present context it draws attention to a chief weakness of the tachyon approach as a way of understanding entanglement.

The reader with some knowledge of special relativity will probably know that one can avoid having tachyons travel backwards in time by positing an absolute reference frame, according to which tachyons always move forwards in time. But special relativity is commonly taken to imply that all reference frames must be equally valid as perspectives for observing physical processes and determining the laws of nature. At the very least, special relativity requires that an absolute reference frame be inaccessible to ordinary perception. Occam's Razor has led most physicists and philosophers to reject such an idea (for a more detailed discussion of the issue, see Maudlin 2002, esp. chap. 3; cf. Berkovitz 2000; for an argument that a universal reference frame does exist, see Craig 2001a, chap. 5; 2001b). The existence of tachyons is therefore regarded as highly unlikely. Of course tachyons are not the only way of accounting for entanglement while minimizing the interpretive shift away from the classical worldview. As we saw in the previous section, several well-known interpretations of the standard formalism of quantum theory also attempt to tame entanglement by saving as much of the classical worldview as possible.

Ingenious attempts such as the ones discussed in the preceding highlight the conservative nature of physics as a disciplinary practice. When an experimentalist discovers a new phenomenon, the theorist's first task is not to begin constructing a new theory but to see whether some previously well-confirmed theory might account for the new discovery. Conservatism must be recognized as an important feature of actual scientific practice, for there is no guarantee that venturing away from existing frameworks will lead to deeper comprehension. On the other hand, although it may be that entanglement will ultimately be understood without appeals to such ideas as nonseparability or relational holism—and of course physicists and philosophers will continue to try out new ideas in a more conservative vein—I am inclined to think that a more serious departure from the classical worldview is warranted. Bohm's early scientific-theoretical efforts (Bohm 1952) to rearrange the quantum formalism for the sake of generating an alternative picture of the physical world produced a significant shift in his own philosophical thought toward holism (Bohm 1980; Bohm and Hiley 1993). More recently, physicists and philosophers have begun to

explore the possibility of interpreting the significance of quantum entanglement from a holist perspective that is not wedded to the particularities of
Bohm's approach.

4   A KEY DISTINCTION

According to Henry Stapp (1989, 154, 159), the phenomenon of quantum
correlations, together with the mathematics of quantum entanglement,
implies "a peculiar kind of macroscopic wholeness—a strange sort of nonseparability of macroscopically separated parts of the universe . . . [an]
intrinsic connectedness of Nature that is alien to the classical notion [derived
from ordinary experience] that spatial separation entails intrinsic separation." Stapp is one of a growing number of physicists and philosophers who
have come to regard quantum entanglement as evidence for a relational
type of holism within physical processes. Consider the remarks of George
Greenstein and Arthur Zajonc (1997, 155), who speak of the particles that
produce quantum correlations as "tangled together into a seamless unity,"
into a relationship that "carries neither messages nor physical causation,
but that always exists." Abner Shimony (1989, 27; italics in original) puts
the same point in more technical language: "The fact that there exist quantum states of two-body systems which cannot be factorized into products
of one-body quantum states means that there is *objective entanglement* of
the two bodies, and hence a kind of holism." As we saw in earlier chapters,
some of the founders of quantum theory spoke in similar terms, albeit at
times disparagingly. More recent attempts to interpret entanglement owe
a significant debt to Jon Jarrett, who introduced an important distinction
into the discussion clarifying what is centrally at stake in the arguments of
Einstein, Podolsky, Rosen, and Bell.

In 1984, more than fifty years after EPR published their argument, Jarrett (1984, 1989) showed that they were right in one respect but wrong in
another (for a helpful analysis of the debate over Jarrett's analysis, see Berkovitz 1998a, 1998b). EPR presumed that the behaviors of the two particles
would necessarily be independent of one another after they had ceased to
interact. This turned out to be wrong, as experimental violations of Bell's
inequality have shown. But EPR were right about the impossibility of superluminal signaling—the physical world is causally "local." The heart of Jarrett's discovery was to find a second condition necessary for deriving Bell's
inequality, which he called "completeness." I will not consider Jarrett's use
of this term but instead focus on his estimation that the failure of completeness belies our commonsense notions of physical individuation: "Although
[two particles are separately] detectable in space-like measurement events,
'they' form a single object, connected in some fundamental way that defies
analysis in terms of distinct, separately existing parts" (1989, 79; cf. Penrose
2005, 584). In other words, Jarrett saw in the failure of completeness what

we have been calling "state nonseparability." To violate Bell's inequality through nonseparability did not imply a contradiction with special relativity because, unlike nonlocality, it did not by itself open the door to superluminal signaling. If one were to give up separability, one could preserve locality even in a Bell-violating scenario. Jarrett's work gave new life to the hope that a "peaceful coexistence" could be negotiated between quantum theory and special relativity (Shimony 1989, 29).

Around the same time Michael Redhead developed an argument distinct from Jarrett's but with similar implications. Redhead suggested the term "robustness" for those causal connections between two particles, including stochastic connections, where the link is impervious to small disturbances. He argued that the correlations between entangled particles are not robust in this particular sense. When particles A and B are entangled, a disturbance registered by A, no matter how small, is immediately felt by B. This led Redhead to conclude that entangled states imply a kind of holism. He identified the probabilities associated with correlations like those discussed in Chapter 3 as "a good candidate for an inherently relational property of the joint two-particle system" (1987, 107; cf. Scarani 2006, 77, who calls them "a pure relative property" of a multiparticle system). He dubbed this nonrobustness a "harmony-at-a-distance," meaning a connection less implicative than actual relativity-violating action-at-a-distance (2001, 156; cf. Penrose 2005, 598–603; for a critique of Redhead's categories, see Maudlin 2002, 150–154). The analogy Redhead employed—of shaking hands with someone who has a "bi-local" hand (shake her hand here, and her "other" hand moves in harmony)—gave his use of the phrase "action-at-a-distance" an epistemological rather than ontological cast, and suggested a holist view of the state of entangled objects.

Don Howard (1989, 228) has also pursued the implications of Jarrett's work, concluding that quantum theory confronts us with a "radical physical holism at odds with our classical intuitions about the individuation of systems and states." Howard calls Jarrett's principle of completeness the principle of "separability," following Einstein's earlier usage, which we employed in Chapter 3. Howard offers a few "philosopher's hints" regarding the character of this new quantum holism, suggesting first that two systems perceived to be at some spatiotemporal distance from one another may nonetheless be spatiotemporally present to each other from the perspective of their interaction. If this were the case, he thinks, we would be unjustified in regarding the two systems as ontologically separate from one another. Howard suggests several possibilities for realizing this idea: overlapping geometries, additional dimensions, wormholes, topological dynamics, and additional manifolds (1989, 251–253; see also Maudlin 2002, 236–239). If space-time can "bend" and "twist," as in general relativity, then perhaps events in space-time that appear to be distantly separated from one another by one particular route are not far apart by another (e.g., through a wormhole). In light of the discovery of entanglement, Howard thinks that it may be possible to assert an ontological holism that sees reality as one nonseparable whole while still doing local physics: "[T]he universe is 'really' one, but once we put a specific question to it, it falls apart quite naturally into apparent parts" (1989, 253).

Paul Teller also sees holist implications in Jarrett's analysis. Noting that the problem of superposition gets "much stranger" when one moves from superpositions describing single-particle systems to those describing multiparticle systems, Teller identifies his view of entanglement with the phrase "relational holism" in accord with his understanding of multiparticle superposition to include peculiarly "relational" properties. These do not simply follow from—they do not "supervene" on—the nonrelational properties of the individual particles (1989, 214). The formalism of quantum theory, he believes, should lead us to reject the idea that all physical properties are either nonrelational properties of individual objects or relational properties that supervene on the former. Teller calls this view of physical properties "particularism" (Howard's "separability"). When it comes to explaining the nonsupervening relational properties of physically entangled states, Teller finds unattractive appeals to action-at-a-distance or tachyonic communication because, as he puts it, "this circle of ideas has no grip" (1989, 215). They also presuppose a particularist point of view, which is what the mathematics of quantum entanglement calls into question. Because the limits imposed by special relativity would appear to be meaningful only when the principle of state separability holds, Teller argues that there is no conflict between the light limit and quantum entanglement when the latter is taken to imply state nonseparability.

5   NONEMERGENT, NONSUPERVENIENT HOLISM

The particular type of relational holism evinced by quantum correlations can be characterized as nonemergent, on the one hand, and nonsupervenient, on the other. Let us examine the issue of nonsupervenience first. As just noted, Teller takes entangled states to be "holistic" in the sense that their properties do not all follow from or supervene on the properties of the subsystems' states (1989; cf. Healey 1989, chap. 5; Esfeld 2001, chap. 8). This view goes against a long line of thinking in the West that can be traced back to Aristotle, who argued in Section 7 of *Categories* that a relation between two things must be explained by specific qualities—"accidents"—that inhere within each thing. The relational property "taller than," for example, must depend entirely or supervene on the height of whatever things are being compared (for a detailed discussion of Aristotle's account of relations, see Brower 2009). I follow Teller in thinking that entangled states can be understood as holistic inasmuch as they have definite properties that do not supervene on the definite properties of the subsystems. I also follow him in thinking of these properties as collective, as "relations." Understood in this way, the relation that carries a definite property value is not a new ontological thing but simply a collective property of the whole. Allow me spell this out in a bit more detail.

Although it is implicit in Teller's discussion, the point that nonsupervening collective properties of a complex quantum system are *not* new kinds of properties applicable only to the whole is never explicitly stated. The standard example of a property that is a "new kind" is the wetness of water. Hydrogen and oxygen do not carry around little bits of wetness by themselves which, when multiplied by Avogadro's number, add up to a tablespoon of wetness. Wetness is an emergent, higher-level property. Appealing to supervenience as introduced earlier, one can say that wetness is a nonsupervenient, emergent property of groups of molecules. On the other hand, the weight of a cup of water is a supervenient, nonemergent property because it depends entirely on the sum of the weight of all the water molecules in the cup. From the perspective of classical physics, these two options appear mutually exclusive and exhaustive. A physical property must either be nonsupervenient (e.g., wetness) and therefore emergent, or supervenient (e.g., weight) and therefore nonemergent. With quantum entanglement we encounter an unexpected hybrid: a nonsupervenient, nonemergent property.

Consider the property of polarization. It is nonemergent in the sense that it can meaningfully be carried by individual photons. The collective properties of a group of photons are conceived of and therefore indissolubly linked to particular identities of the individuals that compose the system *qua* individuals. There is no emergence here in the sense of a higher-level property. However, entanglement is not only nonemergent but also nonsupervenient. The polarization of an entangled photon system is "relational-holistic" in the sense that the system as a whole can carry a definite polarization value through its internal relationships without that value supervening on the definite polarization values carried by individual photons (cf. Mermin 1998). A system of two entangled photons, for example, can have a definite value of polarization only because each individual electron has the capacity to carry the same property. This is what makes polarization a nonemergent property. At the same time, the two-photon system can have a definite polarization value as a whole without implying that either individual photon must also be carrying a definite polarization value.[7] This is what I refer to as the "nonemergent, nonsupervenient holism" of quantum entanglement.

It is especially important to note that although entangled systems have relational holistic properties, with no entangled system is it true that all of its physical properties are relationally instantiated. An electron and a proton, for example, can become entangled through their momenta, but they cannot become entangled through their charges. The electron retains its own distinct charge (negative) even when it and a proton (which has a positive charge) become entangled. The charge of each object is always a nonrelational property related to the enduring identity of the object, whereas the momentum of each object can, through entanglement, become a nonemergent, nonsupervenient relational property as well. Physical objects become entangled through their relational properties even as they retain their individual identities by continuing to possess their own distinct identities (cf. Scarani 2006, 107 n. 16; Barad 2007, chaps. 3 and 4).

Quantum entanglement, therefore, does not imply an emergent, higher-level unity, but rather a coherent plurality—the "relations" as they are manifested by the system and carry its particular state. The entangled relation is no thing in and of itself, but it stands out from the parts inasmuch as they cede to it certain aspects of the task of existing in a particular state (cf. Scarani 2006, 90–91). Because it is nonemergent the relational holism of quantum entanglement is distinct not only from particularism but also from strongly monistic forms of holism. To say that quantum entanglement exists as a ubiquitous feature of the physical world is not to say that beneath the multiplicity of classical existence lies an undifferentiated quantum unity. The qualifier "relational" makes sense only if there really are distinct objects and/or processes to be related (Teller 1989, 222). Quantum states sometimes (often, always?) merge, whereas individual identities do not.

The fact that nonemergent, nonsupervening relations can lead to a violation of Bell's inequality is perhaps not such bizarre news to the antireductionist, but in light of the thoroughly reductionistic character that has typified physics over the past five centuries, it is indeed surprising to find such an idea surfacing within one of its most cherished theories. Paul Jeffries puts it well:

> [Although] we are used to thinking of relational properties as supervening on intrinsic properties (so that "this key fits that lock" supervenes on the intrinsic features of the key and lock taken individually) . . . it may be that the world doesn't work this way at the quantum level; some relational properties may have a "life of their own," having determinate existence independent of any indeterminacy of the intrinsic properties which are the *relata*. (2000, 119)

This insight into the particular and peculiar nature of quantum entanglement informs the theological discussion in the final chapter, but before returning to theology we should examine some of its other theologically relevant features.

6 IMPORTANT FACETS OF ENTANGLEMENT

The idea of relational holism defies our classically honed intuitions about what it means to exist as a discrete, i.e., distinct and separate, entity. Of course relational holism will look different depending on one's broader interpretive framework, but a departure from the classical worldview is unavoidable for any interpretive approach that regards entanglement as a phenomenon worthy of understanding. Read through the lens of relational holism, quantum entanglement has a number of distinctive and interesting characteristics and contours, i.e., its own particular shape. Let us begin by taking up the issue of entanglement's "hiddenness."

As a physical phenomenon, entanglement is doubly hidden from human perception. First, it is hidden macroscopically. It is possible for physicists to discern the presence of entanglement's characteristic features only in the case of subatomic and atomic particles. One ongoing area of research focuses on the challenge of observing larger and larger physical systems in superposition states (O'Connell et al. 2010). It is conceivable that technological advances may someday allow us to see with our own eyes, recognizably and directly, evidence of entanglement in the world, but for now we can only infer it from the behavior of objects that lie at the edge of or beyond the capacities of our own senses. Although entanglement is macroscopic with respect to distance—as noted earlier, it has been demonstrated to exist over distances larger than one hundred kilometers (Ursin et al. 2007)—no object larger than an atomic ion has yet been identifiably entangled in a Bell-type experiment.

The second aspect of entanglement's hiddenness is less obvious. Surprisingly, it turns out that no amount of inspecting the behavior of an individual quantum particle will reveal whether or not it is entangled with some other particle. Recall that one needs to *compare* two *sets* of data, one from each end of multiple runs of a Bell-type experiment, in order to determine whether two collections (ensembles) of particle pairs all prepared in the same initial joint state are entangled with one another—as we saw in the previous chapter. The mathematical description of the particular entangled Bell state we encountered in Chapter 4 shows this clearly. Using Equation 4.12, one can again ask a question considered in Chapter 4, but now with a more hermeneutical agenda in mind: How will a particle in one wing of the Bell-type apparatus shown in Figure 3.5 behave, irrespective of the behavior of its partner in the other wing? Again, to take a specific example, consider the probability that a photon traveling in the left wing will be transmitted when the filter's transmission axis is set to the angle α. Because two of the four joint-outcome probabilities listed in Equation (4.12) contain outcome T_α, the quantum prediction is obtained by summing these two separate probabilities:

$$P_{QM}\left(T_\alpha\right) = P_{QM}\left(T_\alpha \otimes T_\beta\right) + P_{QM}\left(T_\alpha \otimes S_\beta\right) = \frac{1}{2}\sin^2\left(\alpha - \beta\right) + \frac{1}{2}\cos^2\left(\alpha - \beta\right) = \frac{1}{2} \quad 5.1$$

using the fact that $\sin^2(\theta) + \cos^2(\theta) = 1$ for any angle θ. A photon drawn at random from a collection of entangled pairs of photons has a 50 percent chance of being transmitted through the filter, which is exactly what one would predict for an individual photon drawn from a randomly polarized, but unentangled collection of individual photons. Finally, the string of data containing the results of *repeated* measurements in the left wing will also include an apparently random collection of "transmit" and "stop" data points. Nothing about the data pertaining solely to the set of particles on one side of the apparatus reveals that these particles have entangled partners on the other side of the apparatus. In sum, quantum entanglement can only be detected "globally" through the combined application of comparative (synchronic) and statistical (diachronic) analysis, as we noted in Chapter 4. Paul Jeffries refers to this type of hiddenness as "local locality." When we focus our gaze locally, we see only a classical world with some apparent randomness thrown in. When we look globally, however, entanglement comes into view (Jeffries 2000, 121; cf. Scarani 2006, 65).

Entanglement is not only a well-hidden phenomenon, it is also a mostly uncontrollable one. Most importantly, the connection between entangled objects cannot be harnessed for the sake of superluminal signaling. The reason for this, as we discussed at the end of Chapter 4, is that each individual measurement event retains its random character despite the presence of an entangled connection. This negates the possibility of control. A "sender" would need to be able to control the measurement outcome in her own wing of a Bell apparatus in order to force a particular outcome in the distant wing—but this is precisely what the randomness of quantum events prevents her from doing.[8] She might hope to superluminally influence distant measurement outcomes on a statistical basis (i.e., over the course of a large number of runs), but the stochastic character of individual outcomes again steps in to prevent signaling. It is for this reason that entanglement is sometimes referred to as "quasi-causal" or even "noncasual." Recall Redhead's description of quantum correlations as lacking "causal robustness." Entanglement is an intriguing metaphysical puzzle precisely because it appears to be an unmediated relationship that sustains itself among objects which, according to our best available theory of spatiotemporal relations (i.e., Einstein's special theory of relativity), can have no causal relation to one another. From the point of view of a deterministic interpretation of quantum theory, the unmediated character of this relationship is perhaps best understood as an epistemological point because—from the Bohmian perspective, at least—the quantum potential *does* mediate the entangled relationship across space and time. But the quantum potential then becomes the purveyor of global nonseparability. From the indeterministic perspective, on the other hand, the unmediated character of quantum correlations appears to be a genuinely ontological issue because the wavefunction is a description of the joint state of the particles themselves rather than of a distinct entity that mediates their relationship.

Several other aspects of entanglement also deserve mention. First, entanglement is a connection that does not diminish with distance and cannot be obstructed by any known physical barrier. It is also highly discriminating in the sense that only an object's entangled partner feels its influence. In fact, entanglement would appear to transcend the relativistic limits we have come to associate with space and time. The possibility of quantum correlations between spatially (and even space-like) separated objects means that entanglement, understood through the lens of relational holism, does not abide by the normal limits of spatial separation. It also means that entanglement inhabits an ontological realm of ambiguous temporal relations. Recall that special relativity, as commonly interpreted, undermines the idea of an absolute temporal ordering among causally isolated (i.e., space-like separated) objects. If no nonarbitrary temporal ordering of the two measurement events that occur during a single run of a Bell experiment can be given, then might one also speak of entanglement as transcending the normal limits of temporality? Although it is true that one can speak of space-like quantum correlation measurements happening "at the same time"—as one can of any two space-like separated events—this truth will not hold for all reference frames in a relativistic world. Different observers will sometimes disagree about whether the correlations are actually instantaneous. It therefore seems more appropriate to speak of the correlations as *transcending* the normal character of everyday temporal relations.

Is there any justification for thinking of entanglement as a general feature of physical processes? After all, physicists do not tend to speak of "entanglement" generally, but of the "process of entangling" two specific (ensembles of) particles in the laboratory through careful preparation. Perhaps the nominalization grants greater weight to the idea than it deserves. Physicists have never observed entanglement "in the wild," so to speak, but rather entangle and disentangle small numbers of particles in the highly controlled and artificial environments of their laboratories (but see Sarovar et al. 2010; Gómez, Peimbert, and Echevarría 2009). To claim that two photons can become entangled is one thing, but to claim that the physical world as a whole is ubiquitously entangled is quite another. Physicists continue to successfully deploy the tools of classical physics, and engineers can safely ignore quantum entanglement when designing any macroscopic object. Schrödinger hinted at a different view when he wrote that the wavefunctions of two particles "become entangled" by their interaction (1935, 555). From the perspective of quantum theory, particles become entangled whenever they interact. For those who have learned to think and see the world through the language of quantum theory, the most puzzling thing about quantum entanglement is not the idea of living in an entangled world—according to quantum theory, the world should be massively entangled!—but that entanglement isn't a more visible feature of the world. From the quantum perspective, entangled states are the norm, not the exception. Roger Penrose (2005, 591) calls this the "second" mystery of entanglement: "A puzzle that must be faced is the fact that entanglement tends to spread. It would seem that eventually every particle in the universe must become entangled with every other. Or are they already all entangled with each other?" Penrose, voicing what is probably the opinion of most physicists, thinks that the collapse of the wavefunction cuts the bond of entanglement, which would otherwise be ubiquitously present in all physical interactions.

Why, then, do physicists speak of the "accomplishment" of having created entangled particles in the laboratory (and of destroying entanglement in the act of measurement)? A conservative response would be to say that although many forms of entanglement may exist "in the wild," so to speak, most will be unfathomably complex, extremely short-lived, and nigh unto impossible to detect. The role of the "laboratory" in science has always been to offer a specialized environment in which some particular phenomenon can be isolated from the overwhelming complexity of its natural environment. More speculatively, however, one might say that the language of "creating" and "destroying" entanglement in the laboratory is misleading. In order to understand the rationale for this bolder claim, we need to familiarize ourselves with another idea that has been the focus of much debate in recent years, namely, the idea of "decoherence." To a quantum physicist this term signifies the loss of identifiably quantum behavior from a simple quantum object when it interacts with a highly complex, highly disordered object (e.g., its environment). When the idea first surfaced in the 1980s, decoherence garnered significant attention because it appeared to provide a clear and complete solution to the measurement problem on the basis of the *existing* mathematical formalism of quantum theory. There was no need to adjust the math (á la Ghirardi, Rimini, and Weber 1986), yet decoherence appeared to be something more than a new interpretation of the standard formalism (á la Bohm 1952). It seemed to be a previously unnoticed dynamical process—something the pioneers of quantum theory had missed.

The further twist in the story, however, is that most physicists now agree that decoherence does not completely solve the measurement problem. The difficulty, as many have observed (see, for example, Butterfield 2001, 118–120), is that no actual "collapse" takes place when the wavefunction of an object "decoheres" through its interaction with the environment. What happens instead is that the object and the environment become entangled! The individual state of the object, now part of a much more complex but still entangled system, is rapidly driven into a superposition state that is indistinguishable from a classical one. This happens despite the fact that the larger system—the object plus its environment—retains its identifiably quantum character. If one knows where to look, one can still find evidence of the entangled state of the overall system (Greenstein and Zajonc 1997, 206–210). Jeremy Butterfield (2001, 120) has argued that one should therefore "speak not of 'decoherence', but of 'diffusion (into the environment) of coherence'." The counterintuitive result of this diffusion process is that it hides the very behavior that would otherwise mark an individual object as *quantum* rather than classical.[9] If one is inclined to regard decoherence as a "complete solution" to the measurement problem, then entanglement is never actually created or destroyed, only redistributed. Even though most physicists and philosophers want something more by way of a solution to the measurement problem, this new aspect of the theory is now uncontroversially taken to be an important piece of the larger puzzle that is the relation between the classical and quantum worlds. Returning to the question that prompted this digression, one can say that if decoherence is all that happens in measurement then physicists neither increase nor decrease the amount of entanglement in their laboratories—they merely tweak what's already there in such a way as to allow it to manifest itself clearly within their experiments.

One potentially significant implication of this line of thought is that the world of our experience, filled as it is with distinct and separate physical objects, including ourselves, might be continually emerging out of what is otherwise a cosmically and quantumly entangled universe. This is a dizzying thought, to say the least. I hasten to add, however, that at present there is no empirically well-confirmed theory within which it is possible to treat the universe as a quantum object (string theory and other intriguing possibilities remain untestable at present[10]). Still, for a metaphysical perspective that aims to characterize individual identities within a relational ontology, decoherence might just be the "Goldilocks point" that balances relational togetherness against uncountable instances of separation and distinction. Each of the different aspects of entanglement considered in the preceding offers new opportunities for theological reflection and construction, which brings us full circle back to the question of relationality in contemporary theology.

6 Entanglement, Theologically Speaking

[Jesus Christ] is the image of the invisible God, the firstborn of all creation; for in him all things in heaven and on earth were created, things visible and invisible, whether thrones or dominions or rulers or powers—all things have been created through him and for him. He himself is before all things, and in him all things hold together.

—Colossians 1:15–17 (NRSV)

The religious significance of the inorganic is immense, but it is rarely considered by theology. . . . This is one of the reasons why the quantitatively overwhelming realm of the inorganic has had such a strong antireligious impact on many people in the ancient and modern worlds. A "theology of the inorganic" is lacking.

—Paul Tillich (1963, 18)

This final chapter builds upon the relational-holist interpretation of quantum entanglement developed in the previous two chapters by outlining a constructive theological agenda around the notion of entanglement. My central aim here, as I noted in the introduction, is to speak both creatively and responsibly about the God whom Christians claim created *this* world. Is the term "entanglement" ripe for theological appropriation? Can it helpfully be applied to a broad collection of theological concerns? If one of theology's important aims for the early twenty-first century is to reconstrue the world—indeed the cosmos—as "creation," then the discovery of quantum entanglement is an invitation to consider how previous generations of Christian theological speech about the identity of God, the character of the world, and the nature of the God–world relationship might be informed and even transformed by developments in the sciences. My goal is not to construct a physico-theology based on quantum entanglement but to explore how the idea of entanglement might lead to a new and compelling theological portrait of the world and its trinitarian creator.

Here, as in Chapter 2, my primary interlocutors will be Sallie McFague and Wolfhart Pannenberg. McFague has argued (1987, 183; cf. Ruether 1992) that a basic challenge for Christian theology is to avoid characterizing God as essentially unrelated to the world, on the one hand, and as undifferentiated from it, on the other. In her theological account of the world as God's body (1993), which built on her earlier discussion of God as Mother, Lover, and Friend (1987), McFague painted an "imaginative picture" (1987, 92) of a world intimately related to God and a God who cares deeply for the world. I aim similarly to construct a trinitarian image of the relationship between God and the world that builds upon entanglement as the root metaphor. I wish to focus on the possibilities for consonance between various aspects of an explicitly theological characterization of the God–world relation and a (highly interpreted) scientific description of physical reality using quantum theory. For his part, Pannenberg has rightly insisted that theologians should attempt to "think of reality as a whole with the inclusion of nature as a process of a history of God with [creation]" (1993, 112). The philosophical interpretation of entanglement developed in the previous chapter offers theology a unique point of engagement for reflecting on the meaning of *this* physical world, which can contribute to the larger goals of self-enrichment, self-criticism, and continued dialogue with the sciences.

The first two sections of the chapter explore the nature and creative work of the trinitarian God of Christianity through the metaphor of entanglement. I argue that the Trinity is fundamentally entangled "within" Godself and freely entangled "with" creation. The next section then considers several aspects of creation's entangled character. On the one hand, quantum entanglement can be viewed as a theological *vestigium* within creation (see Chapter 1) of God's own relationality. I reserve "entangled with" for the positive senses of God's relation to creation and of creaturely right relations to one another. God becomes freely and lovingly entangled *with* the world through the divinely incarnating act of creation and subsequently through humanly relating acts of blessing, justice, and compassion. On the other hand, much human entanglement can be thought of as an aspect of brokenness and sin. I reserve "entangled in" either for negative instances of finding oneself party to relationships that coercively restrict or harm, or for loving attempts to enter into such relationships purposefully with the goal of effecting restoration. God becomes entangled *in* the world, emblematically on the cross, in response to human acts of violence, indifference, and contempt. Entanglement as a theological metaphor for the God–world relation provides a way of understanding the continued presence of suffering and evil within creation. The fourth and fifth sections consider the efficacy of God's transformative presence in a divinely and quantumly entangled world. If the behavior of an entangled whole is only unambiguously affected *as a whole* by the entangled relationships that exist among its parts, then the entanglement of God with and in the world implies a transforming relationship between God and the world that nonetheless leaves room for the world to continue to be itself, suffering and evil included. In the sixth section I consider the central differences between my own plerotic account of God's power as the basis for the relative freedom of the world to be itself and the route taken by those who espouse a self-limiting kenotic theology of divine power. In the concluding section, I briefly take up the question of whether the idea of entanglement might contribute to a new theological imagination among contemporary Christian communities.

Before turning to the question of divine entanglement, I want to say a word about the role of gendered language in trinitarian discourse. Gendered references to the Trinity ought not to be absolutized, but neither can they easily be set aside. Gendering God is a natural and important part of a tradition that has constructed its images and concepts of divinity in predominantly personal terms. To eliminate gendered language from God-talk would deprive Christian theology of much profound language and affective power. The way toward a solution to the idolatrous masculinization of trinitarian language, which unfortunately permeates much of the history of Christian thought, is not toward the suppression of personal, gendered language when speaking of God but toward the joining of new and old personal characterizations in theology, as well as in prayer and liturgy. Still, moving in this direction will not bring about a complete solution: The exclusivity of personal language *qua* personal language remains. What of God's relation to nonpersonal and inorganic creation?

In the following pages, and out of desire to foreground the vast—nigh unto unfathomable—scope of God's creative act in bringing the universe into existence, I will primarily use McFague's *nonpersonal* triad of Mystery, Physicality, and Mediation (1993, 192) to refer to the members of the Trinity. Although insufficient by themselves to bear the full weight of ideas such as "trinitarian personhood," "inner-trinitarian life," and "being as communion," these terms can nonetheless helpfully direct our gaze beyond what is human to God's relation with *all* creation.[1] Can God, as the divine author of creation, relate and be present to its nonpersonal and even inorganic parts in a mode that honors and embraces the particularities of their own existence? If God is "not less than personal," as Paul Tillich has argued (1951, 245), mustn't this God be capable of enfolding and embracing nonpersonal and inorganic realities as they are, in their otherness vis-à-vis personhood and organicity? Or can God only muster a diminished relationship to the nonpersonal and inorganic world? Nonpersonal metaphors need not replace personal metaphors for the trinitarian God, but their addition can deepen our sense of God as the author of the space and time in which we find ourselves to be alive. What's more, with mainline churches beginning to explore more varied liturgical expressions for God, perhaps there is now greater room within Christian worship for putting multiple images of the divine into conversation with the tradition. Alongside such formulations as "Father, Son, Holy Spirit" and "Mother, Lover, Friend," I want to suggest here that it is important for the sake of all creation and for our own sense of our place within it to make room for non- or transpersonal triads such as McFague's "Mystery, Physicality, Mediation."

Perhaps the most remarkable aspect of any piece of text *qua* text is its ability to take on radically new meaning in new contexts. Malleable in just this way is a word like "entanglement," and infinitely more so the biblical and theological epigraphs that began this chapter. And because the meaning of contexts is no less fluid than the meaning of texts, the emergence of semantic novelty is never a one-way process—interpretation always operates in both directions. Even calling interpretation a "two-way street" fails to capture the numerous and overlapping hermeneutical connections and dependencies that will be present in any robust assertion of meaning. This chapter brings to fruition, albeit in a preliminary way, the hermeneutical agenda introduced in Chapter 1 and developed in Chapters 2 through 5.

McFague has noted (1982, 125) that the powerful theological visions developed by Christianity's great theologians have been based not on metaphors for God or for human beings but on metaphors for the relationship between them. However, she has been a bit too quick to dismiss the natural sciences as a potential source of such metaphors in her estimation that they allow only for "local" causation (1993, 142). The empirical confirmation of quantum entanglement cuts deeply against this estimation. And although McFague is right to insist that scientific and theological models differ in important ways, she overstates the distinction when she argues that scientific models refer to the quantitative dimension of the world and theological models to its qualitative dimension (1982, 106). As a way of understanding the correlated behaviors observed in Bell-type experiments, the relational-holist interpretation of entanglement does not fall neatly on one side or the other of this distinction. Instead, it presents us with a quantitatively based concept that highlights a striking, qualitative feature of the physical world; it expresses relationality in ways that we typically associate with metaphors drawn from the personal realm.[2]

1   AN ENTANGLED GOD—DIVINE RELATIONALITY REVISITED

The first theological task of this final chapter is to examine the suitability of the term "entangled" for characterizing the identity of the Christian God. Naming God as "entangled" will set the stage for subsequent discussions of the divine act of creation and the character of the world itself. We can begin by picking up the discussion from Chapter 2 regarding trinitarian relationality. My rationale for signaling my use of a trinitarian framework at the outset is that the transposition of entanglement into a theological key requires a concrete, traditioned theological context. Without such a context—and I do not mean to say that trinitarianism is the only one available—the meaning and intelligibility of the descriptor "entangled" would follow wholly from its meaning in common speech—which can vary widely—and its use among quantum physicists. This would be highly problematic, for there is no direct connection between the customary meaning of the term or its meaning in quantum physics and its potential for theological application. Although there are many different ways in which one could close the hermeneutical gap, my own approach will be to norm the theological valence of "entanglement" against trinitarian categories. Thus a general trinitarian framework—construed specifically in terms of divine relationality—undergirds my attempt to speak metaphorically of an "entangled God." It is the history and substance of trinitarian thought that secures the theological relevance of quantum entanglement, not the other way around. That said, I do not regard the authority of the trinitarian tradition as absolute. The initial act of construing entanglement in trinitarian terms will invite reconsideration of various aspects of the tradition. Both the conceptual structure of entanglement and its specific meaning in the quantum context will therefore be highly relevant to, albeit not determinative of, the theological discussion.

As noted earlier, a number of contemporary theologians have reclaimed from early Eastern Christian theology the idea that plurality within the Trinity is primary. To prioritize plurality does not mean that one begins with three separate, nonrelational divinities and then incorporates them into a single divine family by adding the idea of eternal relations. The prioritization of plurality within the Godhead can best be understood by thinking of the distinct trinitarian persons as being constituted *qua* persons by their relations to one another. The term that is most commonly used to express the fundamentality of relatedness within the divine life is *perichoresis*. Drawn from the Eastern Greek-speaking theological tradition, the *perichoresis* of the trinitarian persons manifests itself in the fact that they hold their divinity "in relation" by indwelling each other. They are God *together* in and through their relations. The trinitarian tradition posits that God *qua* God is *essentially* a three-personed, i.e., a relational, God.[3] As many have pointed out, this idea of divine personhood is antithetical to the early modern sense of personhood in the West as an independent, autonomous center of activity. For the neo-trinitarians, to be a trinitarian person is "to be in relation" to the other trinitarian persons and, precisely through those relations, to be relationally God.

How can the relational notion of *perichoresis* be understood in light of the relational-holist account of entangled relations given in the previous chapter? Recall that I proposed to interpret quantumly entangled relations as nonemergent, nonsupervening carriers of an entangled system's state. Entangled systems, I argued, carry their states not as individuals but together in their relations among their parts. The most intriguing element of this interpretation is that an entangled system can carry a definite value of a particular property even though its parts lack a definite value for the very same property. In the case of divine being, unlike the case of photons, Christian thought about God has long affirmed that the divine state (or external attributes) and the divine identity (internal attributes or being) coalesce—they are understood to be the same. God does not have a singular unchanging core identity, to which different states can apply. If the trinitarian God is relationality, then the divine relational identity is the divine relational state and vice versa. Thus will we use "entangled" to describe not only the divine state but the divine identity as well; entanglement is the relational seat of divinity within the trinitarian God. The deity of this

God resides not in the persons as *distinct* from one another but within and among the persons as they are *related* to one another, i.e., in the relationality that constitutes them and binds them to one another. Entanglement also expresses the fundamentally active nature of divine relationality. As Pannenberg has noted (1994, 1), the trinitarian God is a God who is active "apart from" or "prior to" the act of creation; the inner-trinitarian relations themselves are divine acts. As a name for the divine nature, entanglement lends itself to an active view by virtue of the fact that it combines the ideas of acting and being; to be "entangled with" is to act by virtue of one's being in a dynamic relation to another.

Here I wish to push the idea that divinity inheres in the relations among the trinitarian persons beyond its traditional formulation that each person is divine. Not only are they not a blended or united pantheon of three independent divinities, they also have no divinity *except* through their relations to one another (cf. Buxton 2005, chap. 5; Simmons 2006, 146–147). To cast mutual indwelling as entanglement provides a novel way of disarming the charge that trinitarian divinity is either the amalgamated divinity of three distinct Gods or the unitary divinity of a single divine subject with three aspects (cf. Jenson 1982, 119). Neither Mystery, nor Physicality, nor Mediation—none of these in and of themselves—should be thought of as divine. One can say that each person or principle is divine as shorthand for the idea that they are constituted by their mutual relations to each other, but for a thoroughgoing relational ontology it can only be a shorthand. Mystery is not in and of itself God. Physicality is not in and of itself God. The Mediation of these two is not in and of itself God. Only through the full set of trinitarian relations can the trinitarian persons be thought of *as divine*, as the trinitarian God. Thus, to say that God's identity is "entangled" is to say that God's divinity is expressed primarily through the relations among the persons. Using language from Chapter 5 and drawing upon the discussion of relational ontology in Chapter 2, we can say that the entangled relations themselves carry the divinity of God because they are prior to the persons. This is a potentially powerful conceptual framework for rearticulating the old idea that divinity is relationality, that being itself is communion (Zizioulas 1985, 2006, 2010).

We saw in previous chapters that some scholars interpret quantum entanglement through the lens of philosophical monism or a physical kind of modalism: The world really is one despite the fact that we experience it as many.[4] Do these perspectives force us to reject the idea of divine entanglement on the grounds that it is in fact a new name for modalism, a new way of characterizing God's nature in which divine plurality is ultimately subordinated to, or flows out of, divine unity? Recall that with quantum entanglement, as I previously interpreted it, the independent behavior and existence of entangled particles is no mere appearance of distinction but a genuine difference, notwithstanding the existence of the holistic relations that fund the particles' behaviors. Quantum entanglement, I have suggested, is not rightly understood as a physical form of modalism or as a weakly differentiated form of monism. It points instead to a relational holism that intertwines the existence of part and whole such that relations among parts bear the state (or in the divine case, the identity) of a system *qua* relations. The relations are *not* the "one" that is always refracted but never actually divided into the "many" of the parts. To speak of divine entanglement is to affirm that genuine differences exist among the trinitarian persons, and that these differences exist by virtue of the divine, perichoretic relations among them.

On the other hand, one can ask whether the divinity of an entangled God is not some fourth thing beyond the persons, a feature or aspect of God in addition to the persons themselves that would effectively render the Trinity a quaternity. In response to this concern we can revisit the point made earlier regarding the identity of God's external and internal attributes. The trinitarian God has not traditionally been understood to "have" characteristics distinct from what the Trinity fundamentally is. In the trinitarian tradition, divine internal and external attribute coalesce— God is what God has. The New Testament speaks not of God *having* love but of God *being* love (I John 4:8). Thus one should not speak of the divine relations as though they were something God has in addition to the divine essence. Again, the trinitarian God *is* these relations. Once again, herein lies the basis for a theologically motivated relational ontology (Polkinghorne 2010b).[5] Recall as well that the relational-holist interpretation of quantum entanglement developed in the previous chapter does not imply the existence of a separate or emergent entangled unity, only a nonemergent, nonsupervenient, coherently entangled plurality, i.e., a "relationality" that is manifested *within* the system. Although entangled relations are not things unto themselves, this does not mean they are nothing. Entangled relations carry the divinity of the persons within the Trinity *as relations*. They are the fundament of divinity, the source of divine personhood.

In Chapter 2 I reviewed Pannenberg's argument that plurality within God does not derive from God's divine oneness but instead that the relations among the trinitarian persons are constitutive for their distinctions *and* for their deity (Pannenberg 1991, 298, 323, 335). Here I want to add that calling God "entangled" is also to agree with Jürgen Moltmann (*contra* Pannenberg) that Christian trinitarianism is *not* best understood as a particular species of monotheism (Moltmann 1993d, 129ff.; but see Peters 1993, 37ff.). Even as Christian trinitarianism must acknowledge with full humility and appreciation its Jewish heritage with regard to its own understanding of God, the trinitarian-relational view of divinity as it has developed over the centuries must also be recognized as a sober letting go of the profound idea of monotheism. Christians do Jews no honor by claiming to have embraced and maintained the view that "the Lord our God, the Lord is one" (Deut. 6:4). More honest, and arguably less supersessionistic, would be to acknowledge that on this particular point a basic difference now separates the two traditions.

In Western theology, the trinitarian Person of the Spirit has traditionally been characterized as the bond between the Person of the Father and the Person of the Son, including their mutual agreement over the Person of the Son's act of turning outside the inner-trinitarian life (Pannenberg 1994, 31). To put the point less anthropocentrically, one could say that in the trinitarian tradition the inner-metaphysical distance within God between Mystery and Physicality is spanned by Mediation. The idea of quantum entanglement works well here because it amounts to a nonpersonal instance of mutual, free agreement. As we saw in the previous chapters the correlated behavior of two entangled particles cannot be attributed to a "common cause" in the classical sense (i.e., their correlated behavior cannot be explained by invoking the possibility of luminal or subluminal communication between the two particles). Each particle is "free" to behave according to its own nature, and yet its entanglement with another particle allows it to act in concert with its entangled partner in a way that cannot be inferred from the presumption of the full separability of individual states. Entanglement is an internal relation that issues in the freedom of particulars. Within an entangled God *all* of the internal divine relations are freely and mutually constituting, and so in a convenient but ultimately inadequate shorthand one can say that the Person of the Spirit is responsible for binding (entangling) the other two persons together. Just as Augustine argued in *De Trinitate* that it was licit to attribute an external act of the whole Trinity to one of its members, so too we can attribute the fundamental relationality *of* the Trinity to one of its members. Within God, Mediation spans the inner-metaphysical distance between Mystery and Physicality—or, to use the more traditional, personalistic terms, the Person of the Spirit eternally entangles the Person of the Father and the Person of the Son. Neither formulation, however, should be taken to imply that the trinitarian God is anything other than relationality itself.

2   GOD'S ENTANGLING ACT OF CREATION

The account of God's "entanglement" given in the previous section prompts us to consider a broader theological narrative about various aspects of God's creative activity. In this section I begin to assemble the main points of this more expansive narrative by examining the entangled, relational character of the divine *act* of creation. Although one could choose to address the entangled character of the world itself before discussing the nature of the divine act that brings it into being—after all, the physical phenomenon of quantum entanglement is the origin of the theological metaphor—this would invite the mistake of thinking about God and the world as entirely separate realities which then must be related to one another by some theological maneuver having little to do with God's initial creative act. Contemporary accounts (e.g., process theology) typically avoid this trap by rejecting the absoluteness of the separation, i.e., by characterizing the God–world relation as fundamental both to the nature of God and to the character of the world (in some cases going beyond the present proposal to reject the idea of an originating "creation" altogether). Taking up the matter of God's entangling act of creation before we consider its own entangled character will help us steer clear of the same trap in the present discussion. If the very act of creation is itself an entangling act, then the fact that an entangled God and an entangled world are in relation to one another becomes theologically basic to God's nature as well as the world's existence. The entangled nature of this relation preserves both the freedom of God in the creative act and the freedom of creation in its own activity.

What might it mean to describe the God–world relation as entangled? I want to argue that the trinitarian perspective must reject any claim of absolute difference between God and the world. The world is not God, but neither is it not-God in an absolute, nondialectical sense.[6] The absoluteness of the difference between God and the world, sometimes associated with the doctrine of *creatio ex nihilo*, must be rejected if the world is to be regarded as the product of a *trinitarian, relational act*. The reason for this lies in the centrality of incarnation, especially in its cosmological dimension, to the Christian Gospel (cf. Rahner 1976, 179; Pannenberg 1994, 21). In the previous section I alluded to Pannenberg's account of the Second Person as the principle of otherness within the Godhead, namely, the Second Person is not only the formative (i.e., patterning) principle of the world in its otherness but also the material (i.e., substantive) principle of its very existence (Pannenberg 1994, 31). Using McFague's notion of the Second Person as Physicality, one can say that the world exists by virtue of its intimate connection to the divine life through the Physicality of God. Alongside traditional accounts of Christ as the Logos, i.e., the idea that is traditionally invoked to underwrite the claim that Christ is the principle of order in creation, a trinitarian account of creation must address the profoundly incarnational idea that the Second Person of the Trinity, in turning outward from the Godhead, is indeed the very stuff of creation. This divine act begins within God, but the universe comes into being alongside—in relation to—God. The world's coming to be is the divine life going beyond itself and its own internal relationality to create an "other" that exists in relation to its own relational ground. Incarnation is the embodiment of physicality.[7]

The point of focusing on the cosmically generative aspect of incarnation is not to claim that the universe emanates from God (*emanatio ex deo*) or is equal to God, but instead to characterize the universe as a created reality that finds its source of existence entirely *within* God.[8] The traditional doctrine of *creatio ex nihilo*, it must be remembered, appeared prior to the development of the doctrine of the Trinity. The *ex nihilo* tradition affirms that God alone is the source of creation, but it fails—and understandably so—to identify the absolute dependence of creation upon its creator in explicitly trinitarian terms. There is no inherent connection between the traditional *nihil* out of which creation comes and the trinitarian nature of God (cf. Pannenberg's critique of Moltmann in Pannenberg 1994, 14). From a trinitarian perspective, however, it is appropriate to locate this *nihil* within God. It can be understood, in other words, as the dialectical *me on* nothingness of the trinitarian persons, separately considered, rather than as the absolute *ouk on* nothingness that has no relation to being (Tillich 1951, 187–188). The outward turn of the Second Person—of Physicality—is a divine turn toward the dialectical nothingness within divine being, but it is also a movement that passes (hovers?) creatively over this dialectical nothingness and extends the inner-relationality of God beyond God. The divinely outstretched hand of Physicality becomes the physicality of our own bodied existence and that of the world around us. As creation, the universe is held fast in relationship with God—which is to say, in its existence—by the Spirit's inward, mediating draw. The world as it exists is not *in* God but *in relationship with* God. From an explicitly trinitarian point of view, the central affirmations of the *creatio ex nihilo* tradition—that the world is good and that it comes from God—can be maintained within a fully trinitarian framework that could more appropriately be called *creatio ex relatione* (creation out of relationality; cf. Ware 2010, 124). Just by virtue of its existence, according to this view, the world has a thoroughly sacramental character. The givenness of its being is the Physicality of God turned outward from the entangled Godhead.

A sacramental view of the world can be found in various contemporary panentheistic theologies of creation (e.g., Peacocke 1993), but it is uniquely motivated within an entangled trinitarian framework. Both the possibility of creation (its otherness) and its actuality (its physicality) are to be found within the divine relational dynamic of Mystery, Physicality, and Mediation. Colin Gunton (1998, 10) was right to insist that trinitarianism can affirm the freedom of the world to be what it is only in light of the idea that the world has its existence in relation to God. How does this bear on the idea of creation as an entangling act? If the Second Person of the Trinity were only the formal principle of distinction within the Godhead, and not also the material principle of creation through the event of incarnation, then there would be no reason to characterize the act of creation as an "entangling" act. The universe would indeed be fundamentally other than God. A cosmic view of the incarnation, however, changes everything in this regard. Although incarnation continues to be an *affecting* moment in which God reveals Godself and God's healing desires for the world to the world, it now becomes an *effecting* moment as well, one that stands at the beginning—albeit theologically rather than chronologically—of the world's very existence (cf. Rahner 1976, 197).[9]

Pannenberg focuses on the category of "distinction" in his account of creation's relation to God (1994, 31), but he also recognizes the importance of the Spirit's mediating role in the act of creation. Just as the eternally and internally relating work of Mediation ensures that Physicality is never separated from Mystery within the Godhead, so the mediating Spirit ensures that creation is never separated from God. In fact, for Pannenberg it is the *cooperative, relational activity* of the Second and Third Persons that must ultimately be said to constitute both the formal principle and the material principle of creation. Physicality and Mediation work *together* through their own mutual entanglement to give the divine act of creation its relational character; an entangled, trinitarian God "relates" the world into existence. This implies that alongside talk of "creaturely distinction" one must also speak of creation in "relational" terms. As Pannenberg puts it, by virtue of the Spirit the Second Person of the Son is "not merely the principle of distinction of the creatures but also of their interrelations in the order of creation" (1994, 32). We will consider this last point in more detail in the following.

Finally, one can connect the entanglement of creation to the entanglement of the cross. Construed as God's embrace of creaturely suffering, the cross is its own entangling moment. It is an event in which the divine relationality manifests its ongoing entanglement with creatures and in their suffering, in the mode not of simple elimination but of relational, transformational, entangled embrace. This mode of entanglement foreshadows the themes that will become dominant when we consider the idea of God's presence in the world in light of the pervasiveness of evil and suffering (Section 4 below). Connecting the entanglement of the divine act of creation to the entanglement of God in and with the world's pain through the event of the crucifixion merges incarnation and cross into a complex whole, through which no neat line can be drawn. The world in which God becomes entangled on the cross due to human sin is the same world that owes its existence to God's transcendent and freely entangling act of creation. Cosmic cradle and earthly cross combine in God's entangling love for the other. But this is as it should be. There must be the closest of all possible connections between incarnation and cross if Jesus's death is not to be regarded merely as a *sui generis* impromptu divine response to human sin.

3   AN ENTANGLED WORLD

Quantum entanglement, understood through the lens of relational holism, can broaden and deepen our awareness of the world's relationality by helping us to appreciate more deeply the unfathomable interdependence, ubiquitous relationality, and immeasurable reciprocity of the universe. McFague has urged theologians to understand God and God's activity in the world "in a fashion that is not just commensurate with an ecological, evolutionary sensibility but intrinsic to it" (1987, 80). I take her to mean that one must strive to envision the world theologically from *within* an ecological, evolutionary, and—lest we forget—nuclear paradigm. Important as an ecological awareness is, we must not rest content with thinking through the theological implications of our kinship to the rest of earthly life. We must also think through the theological implications of the physicality we share with all of creation.

In Chapter 1 I noted that to speak of the trinitarian God is to speak the ideas of "creator" and "creation" in the same breath—all the more so as we come to envision this God through the categories of Mystery, Physicality, and Mediation. Thus far we have considered entanglement as a way of naming God's nature and activity. Having established this basic sense of the metaphor, we can now turn from creator to creation and bring physics back into the discussion more directly. Thus in this section, not only the theological meaning of entanglement but also its physical meaning plays an important role. For if an entangled God entangles the world into existence, then the physical phenomenon of quantum entanglement itself acquires theological significance as a *theological vestigium trinitatis* (see Chapter 1; cf. Buxton 2005, chap. 5). Through a consideration of quantum entanglement we can begin to refigure the physicality of the world as an important aspect of what it means to call the world "creation."

In perhaps the earliest assessment of the theological significance of quantum physics, Robert Russell (1988) suggested that creation's divinely crafted modes of unity and intelligibility might be illuminated by the concept of quantum entanglement (which at the time went by the name of "quantum nonlocality"). Russell characterized God in Tillichian fashion as the "kind of ground of being in which differences need not be contradictions, distinctions need not be isolations, separate entities need not produce alienation, and in which, even at the elementary physical level, distant and simultaneous events need not be ultimately unrelated" (1988, 365–366). He further suggested that quantum entanglement might be construed as a "trace" of the coherence of all events in the divine reality, echoing Henry Margenau's judgment that even at the atomic level nature has a "social" character (quoted in Russell 1988, 351). Russell recognized that the question of nature's character requires not simply a scientific answer, even though scientific considerations must undoubtedly be involved, but a philosophical or theological answer. One prominent view in recent philosophy and theology has been that of "organicism," i.e., that nature as a whole is analogous to a biological organism and must be understood in biological rather than physical terms. But this view no less than mechanism provides a truncated vision of the world. What is physical—what is "inorganic"—is diminished or left out (the latter occurs, for example, in McFague 1997, 22). We human beings are biological creatures, but we nonetheless remain physical creatures in a physical world. And so we must think through the relationality of our own existence and that of all creation not only in biological terms but in physical ones as well.

In his discussion of the world as creation, Pannenberg identifies plurality as the primary mark of its finitude and thus of its created nature (1994, 35).

Although he is interested in the relation between creation's unity and diversity (1994, 61–76), he does not consider whether relationality itself might be another marker of the world's created nature. In the previous chapter I argued that the discovery of quantum entanglement can be interpreted as a relational kind of holism within physical processes. Theologically speaking, this holism can be interpreted as mirroring God's own relationality—both in what quantum entanglement is and in what entangled physical objects do. The sheer existence and relative independence of the world amongst its various parts and in relation to God mirrors the Physicality of God, which is the (personed) principle of differentiation within the Trinity. Likewise, the relationality and interdependence of processes within the world mirrors the Mediation of God, which is the (traditionally less) personed trinitarian principle of fellowship and communion. The particular combination of independence and relationality manifested in the world, especially as it is revealed through quantum entanglement, shuts the door on a fundamentally monistic or strictly pluralistic characterization of nature. The term "universe" derives from the Latin *universum*, which combines *uni* and *versus*, the latter of which implies a "turning outward." Speaking philosophically, a quantumly entangled cosmos is really neither a "universe" nor a "pluriverse." It is a place in which entangled independence-through-relationship is the fundamental characteristic of being. Thus I propose the term *relativerse*. From a trinitarian theological perspective, the relativerse is *creation* in the specific sense that it is the result of the *ad intra* divine entangled relationality turned outward, *ad extra*, toward that relative *nihil* which becomes creation.

In a relativerse the difference between distinction and connection no longer registers as fundamental, despite the useful purposes to which it can be put in attempts to organize and understand the welter of happenings in our various environments. Connection is the basis of distinction in a relativerse, and distinction is the product of connection. The world around us (and within us) manifests this particular type of relationality because God's own entangled being is the ontological basis of its own entangled existence. Pannenberg has characterized the "kingdom of heaven" as that part of creation which is in the immediate and unconcealed presence of God, referring to it as a "multiple unity" (1994, 104). The notion of a relativerse suggests a similar account of the "kingdom (or better, kindom) of earth" as that aspect of creation through which it manifests its own entangled relationality. Imbued with its own relationality, the cosmos mirrors its origin in the trinitarian-relational act of creation. Moltmann's discussion of the eschatological relationship between heaven and earth offers somewhat different imagery, but he spells out their relationship in similarly trinitarian terms: "*heaven* is the chosen dwelling place of *the Father*, but . . . the chosen dwelling of *the Son* is the earth, on which he became a human being, died and rose again, and where he will come in order to fill it with his glory. But then the chosen place of *the Holy Spirit* must be seen in the coming direct bond between heaven and earth in the new creation" (1993b, 162; italics in original).

Referring to the cosmic big bang, McFague has argued that the "oneness of everything at the beginning and hence a common history of kinship is of the most absolute sort imaginable" (1993, 55). Quantum entanglement expands and deepens this view by pointing to a togetherness as profound as the notion of common universal origin. As I have interpreted it, entanglement implies a kind of togetherness among different physical parts of the relativerse that can and does persist across its fourteen-billion-year history and into the present moment. One can point not only to the interconvertability of matter and energy, as McFague does (1993, 47), but also to quantum entanglement as the physical basis for perceiving the cosmos theologically as a place of boundless relationality, a true relativerse. Both the ecological unity of McFague's vision and the holism of quantum entanglement speak to the relationality of the world as it presently exists, with its many diverse parts. Entanglement is an especially potent addition to the ecological view because it bridges the biological and cosmological realms, knowing as it does nothing of boundaries or distances, past or future, life or nonlife.[10]

Now let us look more closely at quantum entanglement from the perspectives of spatiality and temporality. Spatial extension is a necessary condition for creatures to have their existence distinctly from one another. Without it the concept of physical location, as well as the possibility of considering creatures separately from one another, disappears. Temporality, on the other hand, is a necessary condition for creatures to become. Without time, there would be no change in creation and thus no creaturely life. The creative activity of the Spirit supports the duration of forms, in which we have an "inkling of eternity"; the idea of an active divine presence in the world "is hardly possible without time and space" (Pannenberg 1994, 102, 84). Let us take up the issue of temporality first.

Augustine argued long ago (1998, Bk. 11.6) that time is a creature, that God creates not *in* time but *time* itself. Traditionally understood, the consummation of God's temporal act of creation will mark the end of time. Pannenberg and other contemporary theologians have argued, to the contrary, that consummation entails not the end of time but rather the end of separation between past, present, and future (Pannenberg 1994, 95; for a discussion of Pannenberg's temporal holism, see Peters 1993, 168–170; Russell, forthcoming). He identifies the unity of creation with its unified temporal movement toward a common future in God, leaving open the question of whether there are ways in which creation's parts or aspects transcend temporal separation in the here and now. Recall in this regard the temporal ambiguity associated with quantum entanglement. Although entangled physical objects do not lose their temporal identity or location, their relationship to each other cannot be expressed unambiguously in temporal terms if they are space-like separated from one another (this, recall, is because within special relativity space-like separated objects can be assigned no absolute temporal or causal relation to one another; see Chapter 3, note 11). The persistence of quantum correlations under space-like separation implies that entanglement is a nontemporal—or better, a trans-temporal—relation between otherwise temporally situated objects. One might say that quantum entanglement reflects the world's supratemporal origin insofar as it manifests a trans-temporal mode of relationality within its own temporal existence.

The trans-temporal relationality of entanglement does not, from our creaturely perspective, absolutely overcome the distinctions between past, present, and future. Rather, an adequate perception of the unity of creation requires something more than a simply contrastive view of the relation between divine eternity and creaturely temporality. Human consciousness may not be able to perceive this trans-temporal aspect of creation directly, but that does not mean we cannot understand ourselves to be partakers in it (and all the more so if consciousness should turn out to depend in important ways on quantum, perhaps even entangled, considerations). If there is merit to this line of thought, then Pannenberg is not entirely correct to claim that the "unity of life . . . [finds] actualization as a whole only in eternal simultaneity" (1994, 102; cf. 139). Quantum entanglement provides a concrete, this-worldly example of the idea that "the distinction of life's moments in the sequence of time cannot be one of the conditions of finitude as such" (1994, 95). Interpreted in terms of relational holism, entanglement becomes a glimmer of the inner-divine work of Mediation as expressed through the characteristics of the physical world.

In his discussion of the divine life as the basis of spatial distinction, Pannenberg identifies the immensity of God with Samuel Clarke's view of space as undivided extension and the created world *qua* creation with Gottfried Leibniz's view of space as the epitome of relations between separate objects (1994, 89). Quantum entanglement challenges a sharp distinction between divine immensity and creaturely separateness because it forces us to consider created space not simply in terms of the separateness of individual objects but also in terms of undivided extension. As noted earlier, quantum entanglement is a connection that does not depend upon the particularity of spatial relations and which cannot be described simply as the joint workings of two separate and unrelated objects. To use various expressions from previous chapters, quantum entanglement does not allow for state-factorizable mathematical descriptions, it suffers no diminishment with distance, and it cannot be interrupted or blocked by intervening objects. We might therefore regard the spatial nonseparability of quantum entanglement as a reflection of God's unbroken infinity. Pannenberg solves the dichotomy between divine and creaturely space by arguing that the presence of the infinite God in creation is the "*guarantee* of its unity" (1994, 89 n. 229; emphasis mine). Ted Peters argues even more strongly that one "cannot apply [holistic] causation to the universe in the literal sense because . . . the universe is not united in a single causal nexus" (1993, 169). From a classical, Einsteinian perspective, Peters is right. There is no way to speak classically of the relational wholeness of creation. Quantum entanglement, however, provides a possible basis for claiming that the relativerse is a single *trans-causal* nexus. The trinitarian God of this entangled world is not merely the guarantor but also the *source* of its own spatial wholeness.

Like McFague, Pannenberg draws a sharp distinction between the organic and inorganic realms. "[Whereas] the results of the elementary events of natural occurrence seem to remain external to them . . . it is a mark of the organic that it has an inner relation to the future of its own changes and also to its spatial environment" (1994, 33). Here too quantum entanglement challenges the sharpness of the distinction. The inorganic realm has, in a way that is proper to its own mode of existence, something akin to what Pannenberg attributes to the organic realm. Quantum entanglement offers a concrete, physical example of Pannenberg's poetic suggestion that changes in one's own nature resulting from the internalization of external relations yields an "existence on the far side of one's own finitude" (1994, 33). Entanglement, one could say, is a physical rather than personal form of self-transcendence. The difference between organic and inorganic is not that the latter lacks self-transcendence, but that inorganic self-transcendence manifests itself without any form of awareness of itself, whereas organic self-transcendence can also experience itself and, in its more complex forms, even reflect upon itself. I agree with Pannenberg that the evolutionary process ought to be understood in terms of the "increasing intensity of the participation of the creatures in the divine Spirit of life" (1994, 34). But from a cosmic perspective not just the biological but also the physical (inorganic) aspects of creation participate in this "life."

Pannenberg attributes his interest in the broader role of the Spirit in creation to the influence of the writings of Pierre Teilhard de Chardin and Paul Tillich, the latter of whom emphasized the sustaining role of the Spirit in all forms of life (Tillich 1963). Pannenberg appreciates Tillich's unwillingness to restrict the Spirit to the realm of human consciousness, and he finds himself powerfully drawn to the grand vision of Teilhard wherein the Spirit is present in creation by "permeating and activating all the material processes and urging them beyond themselves" (1993, 129).[11] Pannenberg argues, however, that only in self-transcendence can a material being participate in the dynamic of the Spirit—here he has in mind the ecstatic (outwardly directed) nature of human spiritual experience. Peters agrees: "Human beings are not closed monads. We are not like balls on a billiard table that simply bounce around against one another according to the laws of external relations" (1997, 299). What is so important about the experimental confirmation of quantum correlations is that the billiards metaphor is no more adequate for the physical world than it is for human beings. Physical entities too can influence and be influenced by their environment in ways that go beyond the mechanistic picture. Quantum entanglement points to the ecstatic character of creation at the level of physical existence.

4   GOD'S ENTANGLED PRESENCE

We began the chapter by considering divine entanglement. We then revisited quantum entanglement to explore its theological significance for a theological understanding of the world (the relativerse) as creation. In this section we turn back to the idea of God "entangling" the world in order to take up the issue of God's ongoing presence in creation. The distinction between this issue and the issue of God's creative act of relating the world into being is somewhat artificial—having more to do with interest than subject matter—but here we are particularly interested in what might be said about how an entangled God acts transformatively in the world to achieve God's ultimate purposes. The central task of this section is to establish a way of talking about God's ongoing presence in the world that accounts for the apparent independence of creaturely existence without negating God's active presence within creation. I will argue in this section and the next that imaging God and the world as entangled with one another opens up new possibilities for reconciling the idea that God is actively present in the world with the human experience of the world as "Godforsaken" (Moltmann 1993a, 242).

If God were to be present and act in the world by virtue of God's entangled relationship with it, what would that presence look like? If we stick closely for the moment to the analogical aspect of the metaphor—I will pursue it as far as it can go—then we are initially led to the idea that God makes no discernable difference at all to the world's course of events considered on their own. Recall that entanglement does not identifiably change the behavior manifested by a collection of identically prepared quantum objects. It is in fact impossible to tell whether or not a quantum object is entangled simply by examining its own behavior (in isolation from that of its entangled partner). An entangled God does not visibly intervene in the world's processes because "visible intervention" implies an ontological separation between God and the world that does not obtain. Not God's absence or impotence but God's entangled closeness to the world makes divine action invisible and divine intervention impossible. Many who have been willing to follow the theological argument up to this point may find themselves wanting to go no further. If a God who is entangled with the world makes no visible difference in the world and cannot intervene in its processes, hasn't the metaphor clearly outlasted its usefulness? I would ask such readers to suspend judgment for the moment and follow me a little further.[12]

From a theological perspective, I want to suggest that God's entangled relationship with the world makes no *visible* difference to the world in the sense that the entangled God never manifestly steers the world counterfactually in some particular direction through, say, miracles characterized by a general epistemic (i.e., scientific) accessibility. The entangled character of God's relationship to the world grants the world its own causal integrity, and so science sees the world as evolving under its own power. As creatures we are never epistemically *forced* by the entangled presence of God in creation to acknowledge the active and transformative nature of God's presence. Creation really does run according to its own powers. This happens because entanglement funds independence; but it does not mean that an entangled God is absent from the world. It means only that God's action is invisible from our side of the God–world relation by virtue of God's entangled character. God is present in creation in the sense that the divine life of perichoretic love responds to worldly suffering borne of its divinely given freedom by becoming transformatively entangled *in* creation, emblematically so on the cross (cf. Nicolaidis 2010, 106).[13] But even in God's presence on the cross the causal integrity of creation is not diminished. Nothing about Jesus's transformative life and death denies creation its own causal integrity. (I take up the task of rethinking, albeit briefly, important theological concepts such as revelation, resurrection, and salvation in the next section.)

Theological accounts of suffering typically end up saying that God uses or at least allows creation's destructive potential (including our own) to harm and even take life. Within the present view of God and the world as entangled with one another, it would be wrong to say that God uses creation's destructive potential because God's presence in the world is not fundamentally a "using" or "steering" presence. In one sense one can say that God as creator allows the destructive potential to exist. But this is the proximal result of the freedom granted to all of creation by virtue of its entangled relationship with God (in appealing to the freedom of physical processes, I am following Polkinghorne 1989, chap. 5). Humans instantiate this freedom insofar as we find ourselves to be capable of making decisions and acting on them. We also instantiate this freedom insofar as we know ourselves to be the product of an unguided evolutionary process. This is a small part, on the creaturely side, of what it means for God to create "freely."

Allowing the metaphor of entanglement to put pressure on traditional understandings of the God–world relation, as I am doing here, will make it difficult to spell out the details of the relation between evil and God's will. What can be spoken of coherently is the *mode* of God's presence to a world beset by evil. That is to say, a Christian theological account of evil can speak in rich detail about the "how" of God's relation to creation but can name the "why" only in relation to God's desire for relationship with an other who is free. One consequence of focusing on the "how" is that we can avoid the kind of theological machinations that end up pitting theologians against the victims of suffering. We can also affirm God's presence to creation while acknowledging and accepting its divinely given integrity. An entangled God–world relationship implies a world in which physical processes carry their own causal sufficiency. The relativerse is free to run its own course, which is truly its own course, and to unfold in ways that lead to suffering.

5   THE PLEROTIC POWER OF DIVINE ENTANGLEMENT

Although there is not room in this final chapter to explore in detail the connections that might be made among classical Christian themes within a "theology of entanglement," I will indicate briefly some of the beneficial directions in which the discussion might lead. Having identified God as the relational ground of all relationality, and having delimited this divine relationality through a trinitarian appropriation of the metaphor of entanglement, I now want to introduce the idea that the power of God's presence with and in the relativerse is an enabling or empowering presence rather than a dominating or controlling one. Nonetheless, the power of an entangled God can be regarded as "plerotic" (self-filling, self-embracing, self-affirming) in the sense that God grants creation the possibility of relationship and communion through God's entangling presence. Such power is not what we typically think of as plerotic power. The power of the divine presence is far removed from our own ability to control ourselves, others, or our environment. Our own power—which is what we usually think of as plerotic power—is frequently a coercive, self-directed simulacrum of the real thing. When we project this simulacrum onto God we are led to search for particular ways in which God might act in the world to affect its course of events.[14]

But God's entangled, plerotic power is not limited to counterfactual, difference-making action. Instead, it allows the world freely to be and become itself. God's entangled, plerotic power is the power of allowing for the other as other. It is transformative power in the way that quantum entanglement makes a difference to the relationship between entangled objects. Each object is free to behave as if it were unentangled, but the entangled relationship causes the two objects to behave together differently than if there were no entangled relationship between them. They act in synchrony. Likewise, the transformative power of God's entangling presence in creation enables rather than denies the causal integrity of creaturely (including natural) processes, and at the same time transforms the God–world relationship so that the world acts in concert with God even as it continues to act freely on its own. This is the central mystery of entanglement: How can distinct realities act in synchrony without ceasing to act independently? Such is the power of quantum entanglement, and such is the power of an entangled God.

The idea that God freely entangles creation into existence exposes a new theological option for those who experience the world as Godforsaken. My account of God's entangled, plerotic power is motivated by an awareness that at present in human history it has become increasingly hard, to the point of being nearly impossible, for many Christians—including this one—to say of any event in the world: "God did that." I am not insisting that God has less power than the creature, but rather that divine plerotic power has no need of coercion because its source is the very relationality of being (and God said, "Let there be light"). Divine plerotic power creates both the possibility and the actuality of creaturely relationship and freedom. It fosters the general relationality of creation as well as the myriad of particular relationships that constitute the loci of its causal capacities (cf. Zizioulas 1985, 46). This kind of divine power is what we normally associate with the profoundly relational but much abused word "love."

From the beginning of the Jesus movement, Christians have affirmed that the love of God is present in creation because of their sense that the divine desire for right relationship has been revealed in Jesus Christ. In this single and unique person, most clearly in his relationships with others, Christians claim to have encountered the relational ground of existence and to have been shown the meaning of "right" relationship. From a relationalist perspective, the meaning of right relationship must always be arbitrated in community; just as the divinity of the entangled God is present in the relationships between the trinitarian persons, so the revelation of this God can and must be identified through the relationships we form with one another and our surroundings through acts of community. In this sense, incarnation is not only creation but also, in a Christological sense, revelation as well. Raymond Chiao (1999, 16) has suggested provocatively that Jesus Christ can be understood as the "local" manifestation of an entangled God who is both supratemporal and supraspatial. He notes that superposition describes a quantum particle as the materialization at a particular place and time of a superposition that cannot be localized in space or in time. Here Chiao sees a striking analogy to the idea of incarnation. From the visible Christ we learn of the invisible God.

Thanks to its entangled character, divine revelation appears in human history when communities of faith put forward religious hypotheses about the ultimate significance and origin of existence.[15] There is no suspension of creaturely activity in divine revelation, only its continued operation. But this means that actual divine revelation never *commands* assent. Communities of faith make religious hypotheses through their efforts at discernment. From an entangled theological perspective, humans constructing religious hypotheses is just what it means for an entangled God to reveal Godself and be revealed in creation. The most important implication of this view is that God's presence and activity in the world are never beyond debate. When a religious community blesses what it perceives to be good in the world it will inevitably find its judgment contested by others. What we as humans "see" or "observe," i.e., what we find through scientific investigation, is only generations upon generations attempting to understand their lives through communal processes of discernment. What we "perceive," on the other hand, i.e., what Christians, for example, have sensed through their participation in a tradition of communal discernment, is that God's plerotic entangled relationship with creation is revealed in Jesus Christ. Revelation simply cannot be seen, thanks to the world's divinely entangled freedom. In community and through religious traditions, however, revelation can be perceived as communities explore and discern the nature of God's presence in the world by developing a particular religious hypothesis about the ultimate ground and source of its existence. In this sense, revelation is something more than what happens at any given moment. It is what happens over the life of a tradition. Communities test their understanding of the tradition's account of revelation most frequently around the edges but sometimes at the core as well. The perceptions of faith communities and traditions are not unassailable knowledge. They are instead faith-filled convictions about hypotheses that communities have judged to be revelatory. Because of this important difference between seeing (scientifically) and perceiving (theologically), revelation occurs in a divinely entangled world without any abrogation of the full integrity of physical processes. In the realm of reflective self-consciousness (human and perhaps others as well), revelation grants creatures the freedom to do and think as they will. God's entangled, plerotic power lets all of creation be, first by entangling it into existence and then by sustaining it in its entangled freedom.

Communities' perceptions that God has been revealed through their common processes of discernment can grow with time, even to the point where the processes are retroactively identified as religious "traditions." The perceptual weight of a religious tradition brings with it an internal confidence about having received divine revelation, but it never results in the epistemic kind of certainty that would foreclose the possibility of external or even internal debate over its central religious hypotheses. The presence of God's entangled, plerotic love in the world can always be denied. The fruits of religious discernment are always questionable, always debatable. In this sense the revelation of an entangled God is hypothetical in character. The entangled *imago Dei* can be found in our own attempts to perceive and act out right relationships. From a scientific perspective, the *imago* emerges naturally in the freedom of the evolutionary process, but at least one religious hypothesis—the Christian one—entails the divinely given and graced nature of our existence.

Because Christian communities discern God's revelation in Jesus's efforts to enact right relation in his own ministry, we must consider the fact that this ministry came to an end on the cross. As I have interpreted it in the preceding, the cross is the Christian emblem par excellence of God's willingness to love and transformatively embrace the world in its brokenness (Moltmann 1993a). Within a theology of entanglement, however, Jesus's crucifixion need *not* be the center of the concept of salvation. Instead, it can be understood as the expectation-inverting culmination of divine revelation or as an entangling act of divine love that refuses to deny its own nature or the nature of creation in the face of evil. From the perspective of entanglement, the revelatory character of the cross is captured in the notion that God engages suffering transformatively without thereby denying or avoiding the reality of suffering. This would seem like a paradox were it not for our newfound appreciation for the concept of entanglement. God's relationship to the world makes a difference to the world in such a way as to remain completely hidden from "worldly" eyes, a difference that brings its own behavior into accord with God's action without negating in any way its freedom to be itself.

Beyond the cross lies the early church's affirmation of Jesus's resurrection. Here a theology of entanglement will counsel something other than the traditional vision of God breaking into the world to conquer death. If the transformative aspect of God's presence in the world is hidden by virtue of the nature of God's relationship with the world, then this must be true of resurrection as well. In this entangled world death is natural. The untold amount of death prior to our own evolutionary emergence shows this all too clearly. Resurrection can be perceived only from the "global" or theological perspective, not from the "local" or scientific perspective. Recall that the resurrected Jesus was initially perceived only by those with whom he was deeply entangled, his closest followers. In an entangled creation, resurrection can therefore be reenvisioned as the rising of the relational body of Christ in the birth of the church. Bultmann's "emergence of faith in the risen one" (1989, 39–40) was a bold, early attempt to reorient the idea of resurrection in this direction, but his existentialist framework led him to espouse an overly individualistic and fundamentally nonrelational view. As the relational body of Christ, the church becomes entangled with the world through its own attempts to embody and witness the myriad of possibilities for right relationship in a broken world. Here revelation is not only the content of Jesus's life but also the style and method of his ministry. Jesus revealed God through the entangled, relational way in which he taught those around him what it means to live together in awareness of God, which is to say, as the body of Christ.

But whence the theological basis for the picture I am painting, if at its center stands a God whose action can never be seen in the world? Here we run up against the central limitation of using a theological metaphor drawn from the physical sciences. The concept of entanglement knows nothing in its original context of the needs or abilities of consciousness, and so we must accept the responsibility of reshaping it appropriately when applying it to conscious relationships. Rocks and stars have no need of the experience of relationship, but the same is not true of conscious beings. We cannot exist or develop without relationships—all animals know this instinctively if not reflectively from their own experience. This is what pushed us earlier to locate the *imago Dei* in the human desire and need for right relationship. Conscious awareness becomes necessary for the manifestation of our own sense of selves as creatures and, just as importantly, for Jesus's sense of himself as a revelation of God. Above all it is necessary for our sense of the world's relation to God. In none of these respects, however, are the fruits of consciousness immune from criticism or debate.

Readers wanting a tight theological "system" built unflinchingly around the interpretation of quantum entanglement developed in Chapter 5 will no doubt despair at my choice to adjust the metaphor at this point to suit the contours of consciously entangled relationships (I have hitherto avoided any discussion of "divine consciousness" so as to stay focused on the connection between the Physicality of God and its embodiment in creation). Recall, however, that metaphors function by appealing both to an "is" and to an "is not" in the process of comparison. The introduction of creaturely consciousness brings the "is not" of the metaphor of entanglement clearly to the fore. In the end the metaphor must serve our own experience of reality and not the other way around. To demand that all relationships operate in strict analogy with the relational-holist interpretation of quantum entanglement developed in Chapter 5 would be to forget the metaphorical nature of the theological task.

Bringing the cross and incarnation together under the idea of revelation raises the question of salvation, another traditional theological theme that would seem to be imperiled if the entangled character of God's plerotic power in the world precludes God from acting visibly. Two points need to be made. First, from the perspective of God's entangled relation to the world, creation *is* salvation in the sense that God's trinitarian act of entangling the world into existence is an act of salvation *from* the dialectical nothingness of *me on* (see p. 133) as much as it is salvation *to* the dialectical relatedness of existence with a God who is relationality itself. Second, communities of faith can discern right relationships. This typically happens over the life span of traditions rather than individual lives, but even so theological judgments never become internally or externally incontestable. On the Christian hypothesis—which is regarded internally by Christian communities as "revelation"—God enables us to perceive the nature of right relationship through Jesus's words and actions. The world of science sees in Jesus's life and death only human activity, but Christian communities interpret these events through the lens of what they take to be divine revelation.

The metaphor of entanglement can also aid in the perception of God's presence in the world through our own attempts at righting or healing harmful relationships. The entangled God does not step into the world to make all things right—the cross is a clear marker of that, as are the Shoah, the Rwandan genocide, the Indonesian tsunami, Hurricane Katrina, and countless other tragedies that dot the record of human history. But as relational, reflective self-conscious beings, we can embrace our role as God's instruments of love and justice in the world: God makes a perceptible difference when we attempt to make a difference. Note that this view does not entail a consequentialist view of human moral action. Christian ethics need not be about making a clear, visible difference in the world (recall the similarly hidden nature of divine revelation). If it were, the persistent presence of harmful entanglements within human communities, especially when Christians are complicit, would be enough to discredit the Christian hypothesis regarding the efficacy of love. Instead, a Christian ethics of entanglement ought to be about living in right relationship with those around us. The hope for mended relationships must tether itself to the ways in which religious communities perceive God to be transformatively present in the world, not to the visible outcome of our attempts at ethical action.

And what of the ultimate fate of creation, especially the suffering and brokenness that continue to burden our own existence? Recall that Jesus's God, the God of Israel, promises a better future. Jesus perceived this future as already present in paradoxical ways. From the perspective of a theology of entanglement, the transformation perceived and witnessed by Jesus comes with no promise of visibility. Eschatologically speaking, things may end "badly" for us, the Earth, and the relativerse. We now know with near certainty, for example, that life as it has emerged on earth will end when (if not long before) our sun becomes a red giant star and engulfs much of the solar system in several billion years. It is past time for Christians to rethink their denial of any temporal limitations on the future of their relationships with each other and God. The divine promise of transformation does not disappear within a theology of entanglement, but it cannot continue to be an otherworldly promise. The freedom of creation is never abrogated by God's transformative presence, which has to do fundamentally with the *relationship* between God and the world. An entangled God lets creation be its finite self from beginning to end. Indeed, this can be taken as a mark of God's faithfulness to creation. Creation is made new by God through our increasing awareness of our relationship with God. From the perspective of entanglement the promise of transformation is itself transformed into the hope that we envision what is truly possible when we imagine ourselves living into right relationship with each other, with the world, and with God. This vision of right relationship is a divine promise in the sense that it is not fundamentally a lie. It is divine grace in the sense that we really can affect the future for the better. Through this grace we act in concert with God's inner-entangled relationality, even as we are free in this entangled relationship to go our own way.

Much work remains to be done in conceiving and clarifying the relationship of the "personal" connotations of entanglement to its emerging "physical" meaning. Present usage would seem to put a relational notion of sin as entanglement within reach. In at least one instance Pannenberg speaks of living beings "entangled" in the guilt that brings divine judgment (1994, 163). McFague defines sin as a turning away from relationality (1987, 139), which raises the issue of how to relate and distinguish the negative and positive senses of entanglement as I have developed them. If sin is a turning away from right relationality, then can one say that entanglement in destructive relationships denies God's intention for creation to be in right relation with itself and God? Can the prepositional shift from "with" to "in" bear the weight of multiple meanings? The semantic possibilities are intriguing, but only time and usage will tell.

6 PLEROSIS VS. KENOSIS

In a series of related articles on the theological significance of quantum entanglement (1999, 2000, 2006) Ernest Simmons considers several of the themes discussed here. Referring to entanglement as the "physical, metaphorical equivalent of *perichoresis*" (2006, 138), Simmons argues that creation and incarnation can be understood as forms of kenotic entanglement whereby God in Christ goes out of the immanent perichoresis of the Trinity in order to perichoretically embrace the universe. Whereas the Spirit's sanctifying presence in the world is the temporal mode of God's entanglement with the world, the incarnation becomes for Simmons an emblematic intensification of this relationship. He labels his method *analogia relationis*—as the world is interconnected, so God is interconnected with the world. The indeterminacy of quantum systems, he says, reflects the more general freedom given to all of creation through its entangled relation with God. There are clear areas of overlap between my approach and Simmons's, but an important difference relates to my use of entangled *plerosis* rather than entangled *kenosis* to explain the possibility of creaturely freedom.

Although Simmons identifies himself broadly with process theology, he retains the notion of God as creator (in opposition to what would be demanded by strict adherence to the Whiteheadian program). God, according to Simmons, "must first self-limit in order to generate possibility for the creation" (2006, 143). Here is the crux of the kenotic view: God relinquishes the fullness of divine power in the creative act so that creation is not overwhelmed by God's presence. The central difficulty for any kenotic theology that prefers divine self-limitation to metaphysical limitation is twofold. First, the divine act of bringing something into being must itself be a plerotic rather than kenotic act. Even if one supposes that God kenotically "makes room" for creation, the more basic consideration of bringing the world into existence from nothingness requires a plerotic notion of power. As the kenotic theologian John Polkinghorne has noted in his own writing on the subject, "the act of creation, of bringing a world into being and maintaining it in being, is clearly an act of great power to which the puny powers of creatures bear no comparison" (2001a, 90). Simmons, like Polkinghorne, wants to interpret God's self-emptying love as a constraint God freely imposes on Godself for the sake of letting the world be. But what then of the plerotic aspect of creation? And, for that matter, what of the *plerosis* of resurrection and new creation? The necessity of divine plerotic power for conceiving of such events undermines the basic thrust of kenotic theology, i.e., that God self-limits or self-empties for the sake of creaturely integrity and freedom. Second, if God can and does choose to love the world plerotically at crucial points in its history (i.e., in creation, resurrection, and consummation), why does God apparently respond to creaturely suffering and evil in an exclusively kenotic way? Polkinghorne responds to this problem, which is essentially the problem of evil seen through the lens of *kenosis*, by insisting that both *kenosis* and *plerosis* must be affirmed when speaking of God's relation to creation. He acknowledges, however, that this solution is paradoxical and tempers rather than resolves the problem (2001a, 96). Simmons does not explicitly consider the tension between *kenosis* and *plerosis* in his own account of creation, but a similar solution appears necessary. One gathers from this brief assessment that the idea of *kenosis* does not function consistently or effectively across all parts of the Christian narrative. This difficulty is a deep problem for the kenotic program.

One additional feature of Simmons's kenotic account of creation distinguishes it from my plerotic account. Simmons explicitly links entanglement only to the presence of the (economic) Trinity *in* creation, not to the inner life of the (immanent) Trinity. He considers the significance of divine self-limitation only from the historical point of view, in the incarnation and on the cross. In fact, this omission makes sense within a kenotic account of creation. The inner-trinitarian God who must hide the fullness of divine power in the creative act is simply too powerful to be experienced directly by creation. But this puts Simmons in an uncomfortable relationship with Rahner's previously mentioned rule (see p. 27) that the immanent Trinity is the economic Trinity and vice versa (Rahner 1970, 22), which Simmons quotes approvingly elsewhere (2006, 147). If the God who is present in and revealed through the acts of creation, incarnation, and crucifixion is not clearly connected to who God is for Godself—or worse, if God must hide Godself in order to be present to creation without overwhelming it—then *de facto* God as God is for Godself cannot be fully present to the world in the incarnation or on the cross. Simmons appeals to Luther's notion of the hiddenness of God (2006, 143), but the tension between this heritage and (at least one interpretation of) Rahner's rule goes unremarked. The present view by contrast couples divine relationality to an analysis of freedom as it obtains in entangled relationships in order to construe the integrity of natural causes as a sign of God's plerotic presence within creation. Within this approach, the relativerse comes to be *because* of who the entangled God is, not *in spite of* who this God really is apart from any self-imposed limitations. God is present to creatures as God is present to Godself, as entangled relationality.

Finally, an entangled view of the God–world relation lends important theological justification to "methodological naturalism" as the appropriate posture for the sciences in their investigation of reality. Science studies the world of visible, observable relationships and is ill-equipped to consider a God whose entangled relationship with the world precludes empirical observation. To put the point more metaphorically, science confines itself to an empirical or "local" examination of the world, i.e., a view of the world that does not take into account the larger picture that, by the Christian hypothesis, includes God. Theological reflection, insofar as it is rooted in religious community, transcends this limitation by focusing its energies on the "global" God–world relationship. This way of characterizing the distinction between science and theology blocks traditional approaches to natural theology. The debatability of God's presence in creation is not something for theology to mourn or overcome. It is rather a consequence of the fact that God's entangled relationship with the world, by enabling its freedom, supports the notion that it is a closed, causally self-sufficient reality.

7   CAN THIS METAPHOR LIVE?

The possibility of continued theological engagement with quantum ideas such as entanglement depends in large part upon future scientific developments. The vexing issue of the measurement problem has, almost single-handedly, spawned the huge variety of perplexing interpretations of quantum theory that vie for acceptance today. It is conceivable that new empirical or interpretive developments will drastically reshape our understanding of quantum theory in general and entanglement in particular. However, as I said in the introduction, this is a risk to be welcomed by a theology that adopts a fallibilist, hypothetical stance with regard to the assertions it makes about God, humanity, and the world. As quantum entanglement continues to be studied by physicists, new questions will continue to appear on the horizon. What role, if any, do quantum processes and specifically the dynamics of entanglement play in evolution, in the functionality of living organisms, and in self-awareness? What role might they play in future technologies such as quantum computing? Will the conceptual tension between the quantum nonlocal (i.e., instantaneous) view of wave-collapse and the special relativistic principle of locality ever be fully resolved? Are there different kinds of entanglement? Can entanglement manifest itself in highly complex systems? Such questions are interesting in their own right and will, I hope, open new possibilities for a "theology of the inorganic."

Robert Russell has suggested that entanglement is a rich metaphor for the "mysterious and transcendent unity of believers in Christ, and even for our search for a wider ecumenical unity in the global religious perspective." If entanglement were to lead to new ontological insights regarding the traditional dialectic of the one-and-the-many, "reverberations [could] be felt throughout the whole realm of constructive theology" (1988, 367). What possibility is there that entanglement, as a theological metaphor, might someday find a home and role to play in worshiping Christian communities? Of what value is the idea that God is present to the world in ways that make no visible difference to the world? Can this idea help us construct the meaning of our lives and orient ourselves and our actions in the world? The relationship between God and the world is the ultimate referent of any Christian attempt to name the goods we enact, the evils we perpetuate, and those we resist. Do our actions make any difference in an entangled world? Do they reflect the entangled relationship between God and the world? Righting relationships is a basic manifestation of the divine, entangled love that empowers us to love God and our neighbor. Such a love is neither fawning nor clinging but Christic and cruciform; it is a relational "letting be." This kind of love—the love of God's relational and relating being—can only be illuminated by explicitly theological language, not only that used by the theological guild but also liturgical language that attempts to name "in the direction" of love by hypothesizing the ultimate value of things and events as they relate to one another and to God.

Can the metaphor of divine entanglement motivate communities of faith to search more deeply for God and to live more completely into right relationship? The point of the theological life, after all, is not to engage in the abstract exercise of considering the relationality of God at a safe distance but to work toward the mending of all broken and harmful relationships. The entangled creator of this world is closer to us than we could ever be to our spatiotemporal selves; we exist as physical beings only because God is the relationality of Mystery, Physicality, and Mediation. Our bodily existence *is* the Physicality of God turned outward and held gently in the hidden but liberating embrace of God's entangling love. In the midst of this love—even because of it—we and all creation find ourselves endowed with the radical freedom to be ourselves and the world. The discovery of quantum entanglement invites us to experience *all* creaturely relationality as rooted in the relationality of God. The God of the crucified Christ is transformatively at work in the world without intervening in it, so that it might freely become itself through itself. What we do in this world matters supremely because the power and beauty of God's incarnating, entangling gift are manifested, among many other ways, in our attempts to live into right relationship with each other and the rest of creation.

Appendix
Major Theological Works on Relationality since 1980

The theological writings on relationality discussed in Chapter 2 are only a small sample of a much larger body of literature that has appeared over the past several decades. The expanded list given in this appendix is organized chronologically within each theme to help readers track the development of the literature, and to allow those who wish to explore particular topics at greater length identify useful entry points into discussion. Works that address two topics at length are listed under more than one heading, whereas works that cover three or more appear under the first heading, "Relational Theologies and Philosophies." With only a few exceptions, the list provided here is limited to singly or jointly authored book-length works published since 1980. Readers will find a host of relevant journal articles and edited collections in the bibliographies of these texts. See the complete list of references for the full bibliographic details of each text.

ONLINE BIBLIOGRAPHIES

http://www.ctr4process.org/publications/Biblio/Thematic/Relationality.html (focuses on "process" literature).
http://www.opentheism.info/pdf/sanders/bibliography_relationaltheism.pdf (focuses on "open and relational theism" literature).

RELATIONAL THEOLOGIES AND PHILOSOPHIES

Oliver, Harold H. 1981. *A relational metaphysic.*
Heyward, Carter. 1982. *The redemption of God: A theology of mutual relation.*
Suchocki, Marjorie H. 1982 [rev. ed. 1989]. *God, Christ, Church: A practical guide to process theology.*
Oliver, Harold H. 1984. *Relatedness: Essays in metaphysics and theology.*
Zizioulas, John D. 1985. *Being as communion: Studies in personhood and the church.*
Gunton, Colin E. 1991 [2nd ed. 2003]. *The promise of trinitarian theology.*
Pannenberg, Wolfhart. 1988–1993 [ET 1991–1998]. *Systematic theology.*
McFague, Sallie. 1993. *The body of God: An ecological theology.*
Grenz, Stanley J. 1994 [2000]. *Theology for the community of God.*

Edwards, Denis. 1999. *The God of evolution: A trinitarian theology.*
Howell, Nancy R. 2000. *A feminist cosmology: Ecology, solidarity, and metaphysics.*
Keller, Catherine. 2003. *Face of the deep: A theology of becoming.*
Haers, Jacques, and Peter de Mey, eds. 2003. *Theology and conversation: Towards a relational theology.*
Sponheim, Paul R. 2006. *Speaking of God: Relational theology.*
Zizioulas, John D. 2006. *Communion and otherness.*

GOD

Jüngel, Eberhard. 1965 [ET 2001]. *God's being is in becoming: The trinitarian being of God in the theology of Karl Barth.*
Moltmann, Jürgen. 1980 [ET 1993]. *The Trinity and the kingdom: The doctrine of God.*
Jenson, Robert W. 1982. *The triune identity: God according to the gospel.*
Wilson-Kastner, Patricia. 1983. *Faith, feminism, and the Christ.*
Boff, Leonardo. 1986 [ET 1988]. *Trinity and society.*
LaCugna, Catherine Mowry. 1991. *God for us: The Trinity and Christian life.*
Fiddes, Paul S. 1992. *The creative suffering of God.*
Johnson, Elizabeth A. 1992. *She who is: The mystery of God in feminist theological discourse.*
Gunton, Colin E. 1993. *The one, the three, and the many: God, creation, and the culture of modernity.*
Peters, Ted. 1993. *God as Trinity: Relationality and temporality in the divine life.*
Will, James E. 1994. *The universal God: Justice, love, and peace in the global village.*
Jansen, Henry. 1995. *Relationality and the concept of God.*
Torrance, Alan J. 1996. *Persons in communion: An essay on trinitarian description and human participation with special reference to volume one of Karl Barth's Church dogmatics.*
Webb, Stephen H. 1996. *The gifting God: A trinitarian ethics of excess.*
Sawtelle, Roger A. 1997. *The God who relates: An African-American trinitarian theology.*
Cunningham, David S. 1998. *These three are one: The practice of trinitarian theology.*
Sanders, John. 1998. *The God who risks: A theology of divine providence.*
Fox, Patricia. 2001. *God as communion: John Zizioulas, Elizabeth Johnson, and the retrieval of the symbol of the triune God.*
Pinnock, Clark H. 2001. *Most moved mover: A theology of God's openness.*
Pratt, Douglas. 2002. *Relational deity: Hartshorne and Macquarrie on God.*
Shults, F. LeRon. 2005. *Reforming the doctrine of God.*
Baker-Fletcher, Karen. 2006. *Dancing with God: The Trinity from a womanist perspective.*
Bracken, Joseph A. 2008. *God: Three who are one.*
Norris, Thomas J. 2009. *The Trinity, life of God, hope for humanity: Towards a theology of communion.*

CREATION

Moltmann, Jürgen. 1985 [ET 1993]. *God in creation: A new theology of creation and the Spirit of God.*

McFague, Sallie. 1993. *The body of God: An ecological theology.*
Buxton, Graham. 2005. *The Trinity, creation, and pastoral ministry: Imaging the perichoretic God.*
Fretheim, Terence E. 2005. *God and world in the Old Testament: A relational theology of creation.*

ANTHROPOLOGY

Russell, Letty M. 1979. *The future of partnership.*
Russell, Letty M. 1981. *Growth in partnership.*
Smith, Archie. 1982. *The relational self: Ethics & therapy from a Black church perspective.*
Hall, Douglas John. 1986. *Imaging God: Dominion as stewardship.*
Keller, Catherine. 1987. *From a broken web: Separation, sexism, and self.*
Hunt, Mary E. 1989. *Fierce tenderness: A feminist theology of friendship.*
Prokes, Mary Timothy. 1993. *Mutuality: The human image of trinitarian love.*
Suchocki, Marjorie. 1994. *The fall to violence: Original sin in relational theology.*
Volf, Miroslav. 1996. *Exclusion and embrace: A theological exploration of identity, otherness, and reconciliation.*
Grenz, Stanley J. 2001. *The social God and the relational self: A trinitarian theology of the imago Dei.*
Shults, F. LeRon. 2003. *Reforming theological anthropology: After the philosophical turn to relationality.*
Oord, Thomas Jay, and Michael E. Lodahl. 2005. *Relational holiness: Responding to the call of love.*
Oliver, Harold H. 2006. *Metaphysics, theology, and self: Relational essays.*
Cooper-White, Pamela. 2007. *Many voices: Pastoral psychotherapy in relational and theological perspective.*
Battle, Michael. 2009. *Ubuntu: I in you and you in me.*

CHRISTOLOGY

Sullivan, John A. 1987. *Explorations in Christology: The impact of process/relational thought.*
Brock, Rita Nakashima. 1988. *Journeys by heart: A Christology of erotic power.*
Burns, Charlene P. E. 2002. *Divine becoming: Rethinking Jesus and incarnation.*
Shults, F. LeRon. 2008. *Christology and science.*

SOTERIOLOGY

Wheeler, David L. 1989. *A relational view of the atonement: Prolegomenon to a reconstruction of the doctrine.*
Heim, S. Mark. 2001. *The depth of the riches: A trinitarian theology of religious ends.*
Shults, F. LeRon, and Steven J. Sandage. 2003. *The faces of forgiveness: Searching for wholeness and salvation.*

PNEUMATOLOGY

Moltmann, Jürgen. 1985 [ET 1993]. *God in creation: A new theology of creation and the Spirit of God.*
Edwards, Denis. 2004. *Breath of life: A theology of the Creator Spirit.*
Shults, F. LeRon, and Steven J. Sandage. 2006. *Transforming spirituality: Integrating theology and psychology.*

ECCLESIOLOGY

Prokes, Mary Timothy. 1993. *Mutuality: The human image of trinitarian love.*
Volf, Miroslav. 1998. *After our likeness: The church as the image of the Trinity.*
Buxton, Graham. 2005. *The Trinity, creation, and pastoral ministry: Imaging the perichoretic God.*

MINISTRY

Buxton, Graham. 2005. *The Trinity, creation, and pastoral ministry: Imaging the perichoretic God.*
Cooper-White, Pamela. 2007. *Many voices: Pastoral psychotherapy in relational and theological perspective.*

MISSIOLOGY

Roxburgh, Alan. 2000. Rethinking trinitarian missiology. In *Global missiology for the 21st century: The Iguassu dialogue.*

POLITICS AND ETHICS

Smith, Archie. 1982. *The relational self: Ethics & therapy from a Black church perspective.*
Hall, Douglas John. 1986. *Imaging God: Dominion as stewardship.*
Hunt, Mary E. 1989. *Fierce tenderness: A feminist theology of friendship.*
Welch, Sharon D. 1990. *A feminist ethic of risk.*
Webb, Stephen H. 1996. *The gifting God: A trinitarian ethics of excess.*
Sturm, Douglas. 1998. *Solidarity and suffering: Toward a politics of relationality.*
Cho, Hyun-Chul. 2004. *An ecological vision of the world: Toward a Christian ecological theology for our age.*

WORLD RELIGIONS

Sponheim, Paul R. 1993. *Faith and the other: A relational theology.*
Heim, S. Mark. 2001. *The depth of the riches: A trinitarian theology of religious ends.*
Bracken, Joseph A. 2008. *God: Three who are one.*

ESCHATOLOGY

Heim, S. Mark. 2001. *The depth of the riches: A trinitarian theology of religious ends.*

Godzieba, Anthony. 2003. Bodies and persons, resurrected and postmodern: Towards a relational eschatology. In *Theology and conversation: Towards a relational theology.*

Notes

NOTES TO CHAPTER 1

1. Helpful introductions to quantum theory include, in order of least demanding to most demanding: John Polkinghorne (1984, 2002); Valerio Scarani (2006); Nick Herbert (1985); David Z. Albert (1992); and R.I.G. Hughes (1989b). Scarani's (2006) text is perhaps the first popular introduction to focus on entanglement as the theory's defining feature; see especially his discussions on pages xvi and 98, as well as footnote 6 on page 104. A technical introduction to quantum entanglement has been produced by Ingemar Bengtsson and Karol Życzkowski (2008).

2. In *Life Abundant* McFague speaks of fundamental beliefs, those things which we hold most deeply to be true, as "relative absolutes," for we know they are also things about which we might be mistaken. She identifies her own relative absolute as living "to give God glory by loving the world and everything in it" (2000, 29).

3. Ted Peters (1988) has argued *contra* McFague that the solution to the problem of literalization is not to multiply metaphors but to remind ourselves constantly of the "is–is not" character of the established metaphors. In my estimation, this takes too lightly the human penchant for literalization; it also misses the benefits of multiplicity for leveraging the inherent partiality of any single attempt to speak theological truth (cf. Keller and Schneider 2010).

4. Technically, the uncertainty principle is irrelevant to macroscopic physical processes because the constant of proportionality that appears in Heisenberg's uncertainty relation, h (Planck's constant), is incredibly small when expressed in units that make sense for everyday physical objects: $h = 10^{-33}$ joules per Hertz (energy per frequency). That's 0.0000000000000000000 00000000000001 joules per Hertz. Planck's equation, $E = hv$, tells us, for example, that each photon in a monochromatic beam of red light (i.e., with a frequency of $v = 10^{13}$ Hertz, or ten trillion cycles per second) carries only 0.00000000000000000001 (10^{-20}) joules of energy, whereas an ordinary 100 watt lightbulb consumes a relatively enormous 100 joules (10^2) of energy every second.

5. The "mathematical formalism" of a theory is the mathematical structure associated with the theory and is used to generate quantitative predictions. In what follows I will speak frequently of "the quantum formalism" or, more compactly, "the formalism."

6. For a recent example of how quantum theory can be misrepresented, see the 2004 movie *What the Bleep Do We Know!?* The physics of the quantum

world does nothing to warrant the claim that the future can be controlled for personal gain by meditating on desired scenarios.

NOTES TO CHAPTER 2

1. Ian Mevorach (personal communication) has rightfully reminded me that the process by which the doctrine of the Trinity came to represent Christian orthodoxy was frequently ugly. As Mevorach notes, the voices of Jewish Christians and others were silenced and excluded, and the doctrine quickly came to function as a litmus test used to minimize diversity of belief within the church and promote its exclusivist claims regarding salvation. Churches around the globe today still appeal to the traditional doctrine of the Trinity to guard the boundaries of acceptable belief. Although most neo-trinitarian writing is deliberately inclusive if not pluralist, the privileged position from which theologians like myself write owes much to a long-established trinitarian hegemony that has been dismissive of and at times has persecuted non-trinitarian Christians. Neo-trinitarianism, Mevorach points out, "operates within a power-structure that enforces trinitarianism, and [it] takes advantage of this." Neo-trinitarians must be careful not to presume that neo-trinitarianism is the only Western framework capable of registering divine differentiation. Neither should we take for granted that all Christian ears will perceive the "return to the Trinity" in recent theology as the best way of balancing the desire for distinctiveness against the need for cooperation.

2. When referring to the first and second persons of the Trinity, Pannenberg adopts without reservation the traditional, masculine formulations of "Father" and "Son." I let his language stand when quoting him and in discussing his ideas, out of respect for theological difference, although I see nothing in his approach that makes masculine terminology essential to his description of the Trinity. I follow a different route in my own discussion of the issues in Chapter 6.

3. Although one might have thought that Pannenberg would be eager to explore the idea of Christ as *Logos*, he has long been suspicious of this route into Christology (1977, 168), viewing it as an abstraction from the Trinitarian framework. In Volume II of *Systematic Theology*, he connects the idea of *Logos* to the idea of information as a way of speaking about the Son as the principle of structure and order within creation (1994, 112–115; cf. 25–27). I thank Mary Lowe for drawing my attention to this aspect of Pannenberg's thought.

NOTES TO CHAPTER 3

1. Numerous sets of presuppositions have been proposed and used to warrant the derivation of Bell-type inequalities. Disputes continue as to which presuppositions are the most plausible, the most faithful to the spirit of a particular physicist (such as Einstein), the simplest, etc. I employ the perspective developed by Don Howard, who has argued persuasively (1985, 1989, 1990) for the importance of distinguishing between "nonlocality" and "nonseparability" in Einstein's thinking and in current discussions.

2. For simplicity's sake in this chapter I refer to the "standard interpretation" of quantum theory as though the matter were largely settled, when in fact this is manifestly not the case. By "standard" I mean to imply the view—sometimes

also called the "orthodox" view—that broadly affirms the realist intent of scientific theorizing in general, the ontologically indeterministic and indefinite character of at least some physical processes in light of quantum theory in particular (i.e., a probabilistic and realistic interpretation of superposition or the so-called quantum "wavefunction"), and the collapse of one or more of a system's indefinite properties onto a definite value upon some sort of interaction between the system and its environment (typified by measurement). I adopt this strategy only to keep at bay for the moment the complex and controverted landscape of the various competing interpretations of quantum theory. Because the focus of this chapter is the classical viewpoint, such a fiction will suffice for the present. It will need to be deconstructed in Chapters 4 and 5 when we examine quantum theory more carefully and discuss the meaning of quantum entanglement from several different interpretive viewpoints.

3. The "state" of a physical object consists of the values of all of its dynamical properties, such as position, momentum, temperature, etc. Those things that make one type of physical object different from another type, say, the negative charge carried by an electron versus the positive charge carried by a proton, are part of an object's "identity" rather than its "state." To look ahead, quantum theory allows for the superposition of states but not of identities—there is no such thing as a quantum superposition of "being an electron" and "being a proton" (cf. Scarani 2006, 107 n. 16).

4. In fact, Newton puzzled over this feature of his own theory of gravity, according to which distant objects exert forces on one another not locally through contact action but through . . . what? . . . from a distance. It wasn't until Einstein developed his theory of gravity in the early twentieth century that this problem was resolved.

5. Refraction occurs when one throws a stone into the middle of a pond and the resulting ripples slow down as they make their way to the shore through shallower and shallower water; the speed of a wave depends on the characteristics of the medium through which it travels. Diffraction occurs when a wave encounters an obstacle with an edge: it "bends" around the corner created by the obstacle's edge. A good example of interference is the complicated pattern that results from throwing two stones into a pond at the same time but at slightly different places. I will consider the significance of interference for the quantum description of objects in the next chapter.

6. For those interested, the debate over the nature of light continues today, albeit in a more sophisticated manner (cf. Zajonc 2003; Finkelstein 2003; Muthukrishnan, Scully, and Zubairy 2003; Zeilinger et al. 2005; Scarani 2006, chap. 2; Glauber 2007). More importantly, however, violations of Bell's inequality have been shown with other quantum particles such as protons (Lamehi-Rachti and Mittig 1976), neutrons (Hasegawa et al. 2003), and even ions (Rowe et al. 2001). The discrepancy between Bell's prediction and the results obtained in the laboratory cannot be accounted for by the idea that it is a mistake to treat light as composed of particles. Even if one shifts from a particle ontology to a field ontology (as in relativistic quantum field theory), the ideas of property definiteness, state separability, and locality can still be meaningfully considered (Butterfield 2001, 111; Cushing 1989, 15). I employ the particle/object ontology throughout the text, without significant loss of generality on the quantum theoretical side, primarily because it aids visualization.

7. The relation between property definiteness or determinateness, on the one hand, and causal "determinism," on the other, is another potential source of confusion. The latter pertains to the character of the causal connections that underlie the moment-by-moment unfolding of physical processes, whereas

the former pertains only to the character of the physical world at any given moment, i.e., whether it is in definitely one way or the other in any, some, or all respects (cf. Dickson 1998, 14).

8. This section relies heavily on Don Howard's approach to the concept of (non) separability in the context of Bell experiments (Howard 1985, 1989, 1993). I follow Howard's reading of the laboratory violation of Bell-type inequalities as a marker of physical nonseparability, but I arrive at this judgment by a somewhat different route.

9. One further caveat about state separability: It does *not* imply that states A and B are necessarily causally independent from one another. On the contrary, some aspect of the global dynamics of a particular situation might well mean that photon A is never in state $A=T_{90°}$ when photon B is in state $B=T_{90°}$. Even in a fully state-separable world, the state of a physical object might depend on some aspect of its global context (for a discussion of this point, see Howard 1993, section 3). What state separability does imply is that we can resolve the joint state into separate claims about each individual object, taken by itself—which is precisely what we cannot do with a nonseparable state like the one given in Equation 3.1.

10. McMullin also notes that the scholastic theologians accepted Aristotle's principle of contact action, even to the point of denying *divine* action-at-a-distance. God, they said, is omnipresent in creation not by acting at a distance but by being immediately present to all things.

11. An example of two space-like separated spacetime events would be "the sun right now" and "the earth right now." A solar flare that occurs at the former cannot affect the latter. Space-like separated events are, according to special relativity, completely causally isolated from each other. The physical, spatiotemporal sphere of influence of the sun right now is constrained by the speed of light, which is to say that the "the sun right now" can first causally influence the earth 8.3 minutes into the future, i.e., the amount of time it takes light to travel from the sun to the earth. Two such events are said to be "light-like" separated because they can be spanned only by light. Compare this with the two events "the earth right now" and "the sun in one hundred years." These events are "time-like separated" in the sense that the latter is in the spatiotemporal sphere of influence, the so-called "causal future" of the former. What happens on the earth right now can affect the sun at any time greater 8.3 minutes from now. (One hundred years is roughly the amount of time it would take to drive a car going one hundred miles per hour all the way to the sun.)

12. David Mermin claims that Einstein's quip about God not playing dice did not serve him well because history has understood his primary objection to quantum theory to be about its rejection of determinism—the sufficiency of one state of affairs to determine the next. Mermin argues instead that Einstein's unease had primarily to do with an objectively existing reality—with property definiteness (i.e., determinateness) rather than with causal determinism (Mermin 1985, 38). Einstein later admitted the possibility that a quantum particle "obtains a definite numerical value for q (or p) . . . only through the measurement itself" (quoted in Folse 1989, 269), but he continued to object to quantum theory on the basis of his commitment to locality.

13. Physicists use the word "ensemble" to mean a collection of particles all prepared in the same state. The first experiments to clearly demonstrate quantum effects with individual particles were those done with neutrons by Helmut Rauch's group in the 1970s (Rauch et al. 1975; cf. Scarani 2006, chap. 3).

14. The minimum combined degree of indefiniteness of the position Δx and the momentum Δp is set by Heisenberg's uncertainty principle: $\Delta x \times \Delta p \geq h/4\pi$, where h is Planck's famous constant of proportionality.

15. Arthur Fine (1996, 35) has uncovered a letter from Einstein to Schrödinger in which the former confesses that Podolsky actually wrote the paper.
16. The distinction between "locality" and "separability" was not available to Bell.
17. My discussion of the apparatus focuses exclusively on conceptual matters, ignoring a host of miscellaneous and mostly technological details. Readers who enjoy pondering such details would do well to consult the justly famous experiments of Aspect, Dalibard, and Roger (1981, 1982a, 1982b; see also Greenstein and Zajonc 1997, chap. 6) as well as more recent experiments (Weihs et al. 1998; Tittel et al. 1998; Rowe et al. 2001; Zbinden et al. 2001).
18. Today a typical photon source is a unique "nonlinear" crystal (typically made of a borate compound) that absorbs a single incoming photon and emits two new photons with entangled polarizations. Older sources used calcium vapor. Electrons in orbit around the calcium nuclei were excited and thereby caused to emit two entangled photons.
19. "Average preference" is an example of what physicists call an "expectation value," i.e., the average outcome of the same measurement performed repeatedly under the same conditions upon systems prepared in the same state.
20. In each of these scenarios, the left and right polarization filters are oriented differently from one another and are oriented according to a different subset of the three possible angles.
21. A large number of Bell-type inequalities have appeared since Bell first published his own. The inequality derived here is closely related to one that originally appeared in the literature in 1969 and paved the way for the earliest laboratory versions of the EPR–Bell thought experiment. It generalizes Bell's original inequality and is commonly referred to as the "CHSH inequality" in honor of its four co-discoverers: John Clauser, Michael Horne, Abner Shimony, and Richard Holt (Clauser et al. 1969). My own derivation follows the one presented by James Cushing (1994, 193–195). Although Cushing did not identify the origin of his initial algebraic expression, it follows from the CHSH paper. My derivation differs from Cushing's in one important respect: whereas he relied upon two basic physical presuppositions (causal determinism and locality), I rely upon three: property definiteness, state separability, and cause locality (the rationale for this difference appears in note 24 below. For a different physically intuitive derivation of a Bell-type inequality, see Rosenblum and Kuttner (2006, chap. 13).
22. The proof relies upon the factored form of the expression, given by Equation 3.5:

$$Q = x_1(y_1 + y_2) + x_2(y_1 - y_2).$$

Notice that the second half of the factored form of this expression reduces to zero when $y_1 = y_2$, leaving

$$Q = x_1(y_1 + y_2).$$

Each of the remaining variables is equal to ±1, so Q must be ±2. Likewise, the first half of the expression reduces to zero when $y_1 \neq y_2$, leaving

$$Q = x_2(y_1 - y_2).$$

Again, each of the remaining variables is equal to ±1, so Q must still be ±2. These two situations exhaust the possible relations between y_1 and y_2, so in general it must be true that

$$Q = x_1 y_1 + x_1 y_2 + x_2 y_1 - x_2 y_2 = \pm 2.$$

23. An arguably more precise description would be "arbitrary arithmetic combination" because the final term in the expression is subtracted rather than added to the rest. The simpler description "sum" is employed here and in the following on the straightforward grounds that subtracting a positive term is equivalent to adding a negative term.

24. Cushing (1994, 194) does in fact treat the individual and joint terms within single-pair sums as placeholders for outcomes. He uses determinism and locality to argue that, for any actual outcome under some measurement scenario, we are entitled to consider the joint outcome that *would have occurred* if the same pair of photons had been measured under a different scenario. In a deterministic and local world, he says, we would expect that changing only one of the filters' orientations would leave the outcome in the other wing unchanged. Because the same logic applies to all outcomes for all filter orientations, we can assume the same for all outcomes and thus can construct a meaningful algebraic expression out of multiple joint outcomes despite their being mutually exclusive from an experimental point of view. Cushing's argument relies upon the notion of "counterfactual definiteness," i.e., the idea that one can meaningfully compare outcomes that *did take place* with those that *did not take place but could have* (for a discussion of counterfactuality in the context of EPR–Bell arguments, see Redhead 1987, 90–96). I substitute property definiteness and state separability for determinism, not because I find counterfactual definiteness problematic but because relying on determinism to warrant the inequality renders its violation uninteresting for my purposes: One can simply blame the violation on indeterminism, which implies nothing one way or the other about locality. Furthermore, the role of nonseparability is obscured on account of Cushing's failure to distinguish between separability and locality (he did not find this a helpful distinction; personal communication). Of course, if one assumes determinism and accepts Cushing's approach—Cushing himself thought that Bohm's deterministic account of quantum theory had not been given a fair hearing (see Cushing 1994)—then a violation of the inequality points unambiguously to nonlocality. However, in light of the fact that quantum theory cannot be used to devise superluminal signaling devices (e.g., see Shimony 1993, 2:130–139), it is generally accepted that quantum theory does not, strictly speaking, violate locality. I thus construe violations of Bell-type inequalities as evidence for nonseparability rather than nonlocality (more on this in Chapter 5).

25. Schrödinger (1935, 559–560) made this point in his response to the EPR paper with the help of an analogy: If someone does not know ahead of time which of two questions she will be asked but always manages to answer whichever question she *is* asked, it is reasonable to conclude that she must know (in the sense of "have") the answer to both questions ahead of time.

26. One can only ever test a portion of the overall ensemble of photon pairs under any one joint measurement scenario because measuring the entire collection under a single scenario would leave no pairs to be measured under the other scenarios. This unavoidable fact requires an ancillary assumption, namely, that the value for the average preference associated with each joint measurement scenario obtained by testing only a portion of the pairs under a particular joint measurement scenario accurately reflects the value that would have been obtained by testing all pairs under that scenario. An apparatus taking advantage of this "fair-sampling loophole" might be able to respect the principles of the classical worldview and still produce a violation of Bell's inequality. A related concern—the so-called "detection-efficiency loophole"—has to

do with the possibility of real-world measurement devices somehow exploiting their inevitable inefficiencies to skew the result by purely local means, i.e., by conspiratorially failing to detect certain pairs of photons (Fine 1982, 1989a). Bell-type inequalities more sophisticated than the one derived here take such inefficiencies into account, and at least one experiment (Rowe et al. 2001) has already been performed—with ions rather that photons—that achieved perfect detection efficiency, although the experiment left open the "locality loophole" (see the following). Scarani notes with amusement (2006, 87; italics in original) that the detection loophole is "perhaps the first example in the history of physics where the imprecision of the measurements is cited to explain the *perfect agreement* between theory and experiment!"

27. In his 1964 paper, Bell began with simpler expressions, akin to $A_k(\alpha_1)$ and $B_k(\beta_1)$, and *added* a parameter, λ—his famous "hidden variable"—to get expressions akin to $A_k(\alpha_1, \lambda)$ and $B_k(\beta_1, \lambda)$. This maneuver completed the state description and thus rendered the outcome of each measurement independent of the overall configuration of the apparatus. Where Bell introduced an extra parameter into each polarization placeholder, we have eliminated a parameter from each. The two approaches differ primarily in appearance, the rationale being the same in both: Locality guarantees that the filters' orientations have no bearing on the source's production of photons or on each other's measurements during a single run of the experiment.

NOTES TO CHAPTER 4

1. For details about this process, see Greenstein and Zajonc (1997, 135); Grib and Rodrigues (1999, 37–38). The cascade process is an example of what has come to be recognized as one of the signature features of quantum behavior. If a quantum particle traveling through space can take more than one route, and if it is impossible in principle to tell which route the particle travels, then quantum interference effects will appear on the far side of its journey. However, if it is possible in principle to know which route the particle travels—even if no one ever actually checks—then the quantum effects disappear. The situation can even be altered *after* the photon has been emitted from the source and the photon still "knows" how to behave. This phenomenon is referred to as the complementary between "which path" information and quantum interference, and experiments that switch from one scenario to the other after the photon is in flight are known as "delayed-choice" experiments (see, for example, Scully and Druhl 1982).

2. However, as Anton Zeilinger later pointed out (1986, 1), the periodic approach to switching the filters' orientations was less than ideal; only a truly random method for setting orientations could close the locality loophole.

3. The result they obtained violated the CHSH inequality by more than thirty standard deviations.

4. Particles that travel faster than c are in fact allowed, but they can never slow down to c, just as massive particles can never be accelerated to c. No such particles, known as "tachyons" to the theoretical physicists who study them and Star Trek fans alike, have ever been detected. I will have more to say on the plausibility of this explanation for quantum correlations in the next chapter.

5. As in the previous chapter, I use the term "standard interpretation" only to keep at bay for the moment the debates regarding the various competing interpretations of quantum theory. As before, I mean to imply by "standard" a view that broadly affirms the realist intent of scientific theorizing, ontological indeterminism, ontological indefiniteness, i.e., a realistic and probabilistic interpretation of superposition or the so-called quantum "wavefunction,"

and the collapse of the wavefunction upon some sort of interaction between the system and its environment (typified by measurement).

6. Technically, any "linear" sum of states is allowed—no squaring of states, cubing of states, etc., is permitted.

7. This view is variously labeled the "ignorance" or "local hidden variables" interpretation of quantum theory.

8. Because the angle α in Figure 4.4b is greater than $90°$ these two relations are difficult to visualize. The reader can nonetheless confirm from Figure 4.3 that $\sin(\alpha)$ is positive and $\cos(\alpha)$ negative when $90° < \alpha < 180°$, as must be the case if the vectors $\sin(\alpha)T_V$ and $\cos(\alpha)S_V$ are to sum vectorially to T_a within this range.

9. Although determinism was not one of the underlying assumptions of the Bell-type argument given in the previous chapter (nothing about the classical description of polarization properties or the observed outcomes implied the deterministic production of a photon's polarization), Equations 4.5–4.8 provide a helpful window onto quantum theory's signature ability to accommodate the apparent randomness of individual measurement outcomes. The standard interpretation locates this randomness not "externally" at the point of state production, i.e., at the source, but "internally" within the photon itself, i.e., within the multiplicity of basic vectors that characterize its overall state when the photon's polarization state must be expressed as a superposition of the basic polarization vectors associated with the intended measurement. If the expression of a photon's initial polarization state includes more than one basic polarization vector for some measurement, then more than one outcome will be possible when that measurement is performed on the photon. Which of the outcomes actually occurs in any given measurement cannot be predicted; in this sense the standard interpretation says that the results are not just apparently but truly random, i.e., physically underdetermined.

10. A two-dimensional vector and its "right-angled," i.e., orthogonal, projections relate to one another through Pythagoras's theorem, according to which the sum of the squared lengths of a right triangle's two shorter sides must be equal to the squared length of its hypotenuse. One can confirm this visually for S_a by examining the inset of Figure 4.4b and recalling that the length of S_a is 1.

11. Squaring coefficients in the case of joint states works for exactly the same reason it works in the case of individual coefficients. An individual polarization-state vector "lives" in a two-dimensional space and can be "projected" onto some set of basic polarization vectors corresponding to a possible polarization measurement (i.e., a possible measurement angle). By comparison, a joint-polarization vector lives in a four-dimensional space and can be projected onto some set of four basic joint-polarization vectors (corresponding to the outcomes TT, TS, ST, SS under some measurement scenario). The relationship between the length of the joint-state vector and the length of the scaled basic vectors is impossible to visualize, but this relationship is simply a generalized "four-dimensional" version of the Pythagorean theorem: $a^2 + b^2 + c^2 + d^2 = e^2$, where e is the length of the original joint-state vector.

12. For a discussion of the singlet state in the case of intrinsic angular momentum or "spin," see Bohm (1989, chap. 17, 614–623).

NOTES TO CHAPTER 5

1. See the collection of papers edited by J.A. Wheeler and Wojciech Zurek (1983) for many of the early papers. For a recent scholarly summary of the significance of Bell's work on entanglement, see the entry "Bell's Theorem" by Abner Shimony in the online *Stanford Encyclopedia of Philosophy* (2004).

To date, popular accounts of the history of entanglement include works by Amir Aczel (2002), Brian Clegg (2006), and Louisa Gilder (2008). Although all are quite helpful for the uninitiated reader, none pays any attention to the recent philosophical debates. Aczel focuses more than the others on the characters who brought Bell's work into the laboratory; Clegg's explanations of the physics are the sharpest; and Gilder provides an elaborate, painstakingly researched, novel-like reconstruction of the thoughts and words of those who made entanglement the *sine qua non* of quantum theory, from the 1920s to the present. See also Barad (2007).

2. The wavefunction is a concise mathematical way of expressing a superposition state that contains anywhere from two to an infinite number of possible outcomes. Take classical momentum, for example. It is a continuous rather than discrete variable, i.e., its value upon measurement always ranges over an infinite number of possibilities. (This follows from the fact that the distance on the number-line between any two values of momentum can always be divided into an infinite number of segments, no matter how small the distance is.) Momentum must therefore be represented by the integral of a continuous function rather than by the arithmetic sum (i.e., a simple superposition) of a finite number of terms.

3. The quantum potential acts not on the basis of its strength but on the basis of its form, which depends upon the global configuration of all particles in the universe at any given instant. Any change in the global configuration of matter results in an instantaneous change in the quantum potential that leads in turn to local changes in a given particle's behavior (see Greenstein and Zajonc 1997, 144–148; Dickson 1998; Cushing 2001; Russell 2001, appendix; Maudlin 2002, 116–121).

4. Bohm's approach requires, at least in principle, that all spatial parts of the universe have a common "now." This dependence on the notion of absolute "simultaneity" has led many to see a conflict between the Bohmian interpretation and Einstein's special theory of relativity, the latter of which is commonly taken to undermine the existence of any absolute "now" (see, for example, Redhead 2001). Strictly speaking, however, special relativity rules out only nonarbitrary judgments about what constitutes the absolute now. Einstein's general theory of relativity allows one to reassert a cosmic (absolute?) "now" that is hooked to the reference frame of the cosmic microwave background radiation left over from the big bang.

5. But of course the choice among different metaphysical and theological options is never forced by scientific data or theories. The lack of clear entailment from the former to the latter means only that philosophers and theologians must take full responsibility for their claims. This is not to say that the enterprise of science is less hermeneutical than that of philosophy or theology, but only to acknowledge that reality has provided scientists with the least wiggle room as they go around the hermeneutical circle from experimenting to theorizing and back again.

6. The possibility of mutual inconsistency follows from the fact that, according to special relativity, traveling faster than the speed of light entails the possibility of backwards-in-time travel and therefore backwards-in-time causation. Tachyons travel faster than light, and so they will inevitably be seen to travel backwards in time from the perspective of some subluminal observer, i.e., some subluminal reference frame. The causal paradoxes that can appear when backwards-in-time motion is allowed have led philosophers to regard such motion as (1) impossible or (2) incapable of making any difference to the flow of history. A standard example of how altering the flow of history leads to a causal paradox involves a person who attempts to travel back in time to

prevent his own existence, say, by preventing his parents from ever meeting one another. The logician intervenes at this point in the argument and notes that the person must necessarily fail in his quest because his presence is both logically and materially inconsistent with his succeeding.

7. Here I show my preference for the indefiniteness associated with the standard interpretation over the nonclassical definiteness of Bohm's approach. To understand the point, consider a fanciful analogy: Albert and Betty are walking in quantumland in opposite directions along a sidewalk running north and south, and yet it also happens to be the case that neither is definitely walking north or south. If this were the case, then the property of the whole, "walking in opposite directions," could not be said to supervene on the definite properties of Albert and Betty as individuals, such as "Albert is walking north" and "Betty is walking south."

8. A clear example of this is the "perfect" anti-correlations of same-axis measurements, which are not any more controllable for their being perfect. Each joint outcome still appears randomly, which blocks any form of control.

9. Technically speaking, the source of the masking effect is "the loss of phase coherence," hence the shortened term "decoherence." The mathematical tool needed to see how decoherence emerges from the standard quantum formalism is called the "density matrix." An introduction to this tool lies beyond the scope of this work, but the interested reader is encouraged to study the succinct presentation given by Greg Egan (2002). For a more detailed account, see Zurek (2002).

10. But see the argument by Mark Van Raamsdonk (2010) that space-time itself is the result of entanglement. This essay won First Award in the 2010 Essay Competition of the Gravity Research Foundation.

NOTES TO CHAPTER 6

1. One finds this broader connection between incarnation and physicality in Maximus the Confessor: "The Logos of God, who is God, wills always and in everything the realization of the mystery of his embodiment" (quoted in Zizioulas 2010, 154).

2. In one of her earlier works, McFague commented briefly that within quantum theory physical entities "exist only in relationships . . . the process or network of relationship is more basic than the 'substantial individual'" (1993, 125). Although McFague had in mind wave-particle duality, this is an apt description of quantum entanglement.

3. For this reason I avoid descriptors such as "three-in-one" and "triune," which suggest that a more fundamental unity lies beneath the divine plurality. More on this later.

4. See, e.g., Howard (1989, 253); David Bohm's holistic metaphysical and theological views of entanglement have their own particular character, which I cannot elaborate here. For an introduction to Bohm's views, see the work by Bohm and David Hiley (1993) and several of the articles in vol. 20, no. 2 (1985) of the journal *Zygon*. Kevin Sharpe (1993) provides an extended philosophical and theological analysis of Bohm's oeuvre.

5. The essays collected in *The Trinity and an Entangled World* (Polkinghorne 2010b) focus explicitly on the matter of "relational ontology." Unfortunately, despite the book's title and two contributions on entanglement by physicists, none of the theological essays apart from Polkinghorne's opening essay addresses the concept.

6. Tillich's metaphorical characterization of God as the "ground of being" (1951, 235) might be similarly construed as implying a world that is neither God nor not-God (Chi Sang Woo, personal communication).

7. Zizioulas (2010, 150), invoking Dionysius the Areopagite and Maximus the Confessor, articulates a similar vision: "God as *eros* and *agape* (there is no difference between the two terms) moves toward creation, causing at the same time a reciprocal movement toward himself from the side of creation."

8. John Polkinghorne (2000, 89–95) has argued for an "eschatological" panentheism, whereas taking the cosmic dimension of incarnation seriously amounts to embracing a "primordial" panentheism instead.

9. I thank Robert Russell and Wesley Wildman for drawing my attention to Rahner's similar view of the relation between creation and incarnation.

10. Recall the discussion of entanglement's cosmic ubiquity at the end of Chapter 5. There is already reason to believe that quantum entanglement operates at a cosmic scale (Gómez, Peimbert, and Echevarría 2009), and there are strong hints that quantum characteristics play a nontrivial role in the evolution and inner workings of life (see Russell 2001, 293–328; Davies 2009).

11. Stanley Grenz insightfully argued (1999, 164) that the scientific foundation of Pannenberg's "field pneumatology" (see Pannenberg 1994, 79ff.) lies in the "biologically based idea that life is essentially ecstatic."

12. Graham Buxton prefers to use quantum entanglement nondialectically and sees in it a trace of the perichoretic nature of God in the physical world, whereas I use quantum entanglement and *perichoresis* dialectically, interpreting each in terms of the other and starting the hermeneutical process from the vantage point of the trinitarian tradition. I accept Buxton's critique of my earlier view that perichoresis and entanglement ought to be regarded as synonyms, but I suspect he would still be concerned that the present proposal blurs "the radical distinction between finite and infinite" (Buxton 2005, 207). My sense of the matter is that regardless of the type of radical distinction a trinitarian posits between God and creation, the distinction itself must still be grounded "materially" *in* God if one is not to fall back into a *nontrinitarian* account of creation. The idea of the Second Person of the Trinity as Physicality and thus as the material principle of creation, combined with a cosmic interpretation of incarnation, opens the door to this approach.

13. It is possible that future work on quantum entanglement among different types of physical objects will afford further nuances with regard to the metaphorical characterization of God and the world being in an asymmetric, divinely entangled relation. There is no requirement within quantum theory that entangled entities be of the same type; any two physical objects become quantumly entangled, at least momentarily, when they interact.

14. This is the traditional problem of "special divine action," which was taken up recently by an interdisciplinary group of scholars under the auspices of The Center for Theology and the Natural Sciences and The Vatican Observatory (see Russell, Stoeger, and Coyne 1988; Russell, Murphy, and Peacocke 1995; Russell, Murphy, and Isham 1996; Russell, Stoeger, and Ayala 1998; Russell et al. 1999, 2001; Russell, Murphy, and Stoeger 2008).

15. Religious "hypotheses" are not abductively inferable from empirical data. Rather, they are theological possibilities claimed and embraced by communities of faith for the sake of argument and action but with the ever-present assumption of revisability.

References

Aczel, Amir D. 2002. *Entanglement: The greatest mystery in physics.* New York: Four Walls Eight Windows.

Aharonov, Yakir, and David Bohm. 1959. Significance of electromagnetic potentials in the quantum theory. *Physical Review* 115: 485–491.

Albert, David Z. 1992. *Quantum mechanics and experience.* Cambridge, MA: Harvard University Press.

Albert, David Z., and Barry Löwer. 1988. Interpreting the many-worlds interpretation. *Synthese* 77: 195–213.

———. 1989. Two no-collapse interpretations of quantum mechanics. *Nous* 12: 121–138.

Aspect, Alain, Jean Dalibard, and Gérard Roger. 1981. Experimental tests of realistic local theories via Bell's theorem. *Physical Review Letters* 47: 460–463.

———. 1982a. Experimental realization of Einstein-Podolsky-Rosen-Bohm gedanken experiment: A new violation of Bell's inequalities. *Physical Review Letters* 49: 91–94.

———. 1982b. Experimental test of Bell's inequalities using time-varying analyzers. *Physical Review Letters* 49 (25): 1804–1807.

Augustine. 1963. *On the Trinity.* Trans. S. McKenna. Washington, DC: Catholic University of America.

———. 1998. *The city of God against the pagans.* Trans. R.W. Dyson. Cambridge: Cambridge University Press.

Baker-Fletcher, Karen. 2006. *Dancing with God: The Trinity from a womanist perspective.* St. Louis, MO: Chalice.

Barad, Karen. 2007. *Meeting the universe halfway: Quantum physics and the entanglement of matter and meaning.* Durham, NC: Duke University Press.

Barber, Bruce, and David Neville, eds. 2005. *Theodicy and eschatology.* Adelaide: Australian Theological Forum.

Barbour, Ian G. 1971. *Issues in science and religion.* New York: Harper and Row.

———. 1974. *Myths, models, and paradigms.* San Francisco: Harper and Row.

———. 1997. *Religion and science: Historical and contemporary issues.* Rev. and exp. ed. San Francisco: HarperSanFrancisco.

———. 2002. *Nature, human nature, and God.* Minneapolis, MN: Fortress.

Barrett, Jonathan, Daniel Collins, Lucien Hardy, Adrian Kent, and Sandu Popescu. 2002. Quantum nonlocality, Bell inequalities, and the memory loophole. *Physical Review A* 66 (4): 042111.

Barth, Karl. 1975. *The doctrine of the Word of God.* 2nd ed. Trans. G.T. Thomson and H. Knight. Ed. G. W. Bromiley and T.F. Torrance. Vol. 1.1 of *Church Dogmatics.* Edinburgh: T and T Clark.

Battle, Michael. 2009. *Ubuntu: I in you and you in me.* New York: Seabury.

Belenky, Mary Field, Blythe McVicker Clinchy, Nancy Rule Goldberger, and Jill Tarule. 1986. *Women's ways of knowing: The development of self, voice, and mind.* New York: Basic Books.

Bell, John S. 1964. On the Einstein Podolsky Rosen paradox. *Physics* 1: 195–200.

———. 1982. Quantum mechanics for cosmologists. In *Quantum gravity 2*, ed. C.J. Isham, R. Penrose, and D. Schiama. Oxford: Clarendon.

———. 1983. On the Einstein Podolsky Rosen paradox. In *Quantum theory and measurement*, ed. J.A. Wheeler and W.H. Zurek. Princeton, NJ: Princeton University Press.

Bengtsson, Ingemar, and Karol Życzkowski. 2008. *Geometry of quantum states: An introduction to quantum entanglement.* Rev. ed. Cambridge: Cambridge University Press.

Berkovitz, Joseph. 1998a. Aspects of quantum non-locality—I: Superluminal signalling, action-at-a-distance, non-separability and holism. *Studies in the History and Philosophy of Modern Physics* 29B (2): 183–222.

———. 1998b. Aspects of quantum non-locality—II: Superluminal causation and relativity. *Studies in the History and Philosophy of Modern Physics* 29B (4): 509–545.

———. 2000. The nature of causality in quantum phenomena. *Theoria* 15 (37): 87–122.

Birch, Charles. 1972. Participatory evolution: The drive of creation. *Journal of the American Academy of Religion* 40 (2): 147–163.

Boff, Leonardo. 1988. *Trinity and society.* Trans. P. Burns. Maryknoll, NY: Orbis.

Bohm, David. 1952. A suggested interpretation of the quantum theory in terms of "hidden variables" (I and II). *Physical Review* 85: 166–193.

———. 1980. *Wholeness and the implicate order.* New York: Routledge and Kegan Paul.

———. 1989. *Quantum theory.* New York: Dover.

Bohm, David, and Yakir Aharonov. 1957. Discussion of experimental proof of the paradox of Einstein, Rosen and Podolsky. *Physical Review* 108: 1070–1076.

Bohm, David, and Basil J. Hiley. 1993. *The undivided universe: An ontological interpretation of quantum theory.* New York: Routledge.

Bohr, Niels. 1983. Can quantum-mechanical description of physical reality be considered complete? In *Quantum theory and measurement*, ed. J.A. Wheeler and W.H. Zurek. Princeton, NJ: Princeton University Press.

———. 1987. Quantum physics and philosophy—causality and complementarity. In *Essays 1958–1962 on atomic physics and human knowledge.* Vol. 3 of *The philosophical writings of Niels Bohr.* Woodbridge, CT: Ox Bow.

Born, Max, ed. 1971. *The Born-Einstein letters: Correspondence between Albert Einstein and Max and Hedwig Born from 1916 to 1955 with commentaries by Max Born.* Trans. I. Born. New York: Walker.

Bouma-Prediger, Steven. 2001. *For the beauty of the earth: A Christian vision for creation care.* Grand Rapids, MI: Baker Academic.

Bracken, Joseph A. 1984. Subsistent relation: Mediating concept for a new synthesis? *Journal of Religion* 64 (2): 188–204.

———. 2008. *God: Three who are one.* Collegeville, MN: Liturgical Press.

Brock, Rita Nakashima. 1988. *Journeys by heart: A Christology of erotic power.* New York: Crossroad.

Brower, Jeffrey. 2009. Medieval theories of relations. *Stanford Encyclopedia of Philosophy.* http://plato.stanford.edu/entries/relations-medieval/ (accessed August 25, 2010).

Brown, Julian. 2000. *The quest for the quantum computer.* New York: Simon and Schuster.

Bub, Jeffrey. 2010. The entangled world: How can it be like that? In *The Trinity and an entangled world: Relationality in physical science and theology*, ed. J. Polkinghorne. Grand Rapids, MI: Eerdmans.

Bultmann, Rudolf. 1989. *New Testament and mythology and other basic writings*. Trans. S.M. Ogden. Philadelphia: Fortress.

Burns, Charlene P.E. 2002. *Divine becoming: Rethinking Jesus and incarnation*. Minneapolis, MN: Fortress.

Butterfield, Jeremy. 2001. Some worlds of quantum theory. In *Quantum mechanics: Scientific perspectives on divine action*, ed. R.J. Russell, P. Clayton, K. Wegter-McNelly, and J. Polkinghorne. Vatican City State: Vatican Observatory/ Center for Theology and the Natural Sciences.

Buxton, Graham. 2005. *The Trinity, creation, and pastoral ministry: Imaging the perichoretic God*. Milton Keynes: Paternoster.

Cabello, Adán. 1998. Ladder proof of nonlocality without inequalities and without probabilities. *Physical Review A* 58 (3): 1687–1693.

Carson, Rachel. 1994. *Silent spring*. Boston: Houghton Mifflin.

Cartwright, Nancy. 1983. *How the laws of physics lie*. Oxford: Oxford University Press.

Caton, Peggy. 1987. Gender relations: A cross-cultural dilemma. In *Equal circles*, ed. P. Caton. Los Angeles: Kalimat.

Chiao, Raymond Y. 1999. The quantum wave of faith. *Science & Spirit* 10 (1): 16.
———. 2004. Conceptual tensions between quantum mechanics and general relativity: Are there experimental consequences? In *Science and ultimate reality: Quantum theory, cosmology, and complexity*, ed. J.D. Barrow, P. Davies, and C.L. Harper Jr. Cambridge: Cambridge University Press.

Cho, Hyun-Chul. 2004. *An ecological vision of the world: Toward a Christian ecological theology for our age*. Rome: Pontificia Universita Gregoriana.

Chung, Paul. 2002. *Martin Luther and Buddhism: Aesthetics of suffering*. Eugene, OR: Wipf and Stock.

Clarke, W. Norris. 1973. A new look at the immutability of God. In *God knowable and unknowable*, ed. R.J. Roth. New York: Fordham University Press.

Clauser, John F., Michael A. Horne, Abner Shimony, and Richard A. Holt. 1969. Proposed experiment to test local hidden-variables theories. *Physical Review Letters* 23: 880–884.

Clayton, Philip. 1989. *Explanation from physics to theology: An essay in rationality and religion*. New Haven, CT: Yale University Press.

Clegg, Brian. 2006. *The God effect: Quantum entanglement, science's strangest phenomenon*. New York: St. Martin's.

Coakley, Sarah. 2009. Is there a future for gender and theology? On gender, contemplation, and the systematic task. *Criterion* 47 (1): 2–11.
———. 2010. Afterword: "relational ontology," Trinity, and science. In *The Trinity and an entangled world: Relationality in physical science and theology*, ed. J. Polkinghorne. Grand Rapids, MI: Eerdmans.

Cobb, John B. 1992. *Sustainability: Economics, ecology, and justice*. Maryknoll, NY: Orbis.

Cobb, John B., Jr. 1984. Overcoming reductionism. In *Existence and actuality: Conversations with Charles Hartshorne*, ed. J.B. Cobb Jr. and F.I. Gamwell. Chicago: University of Chicago Press.
———. 1986. Christology in "process-relational" perspective. *Word and Spirit* 8: 79–94.

Conner, David E. 2006. Quantum non-locality as an indication of theological transcendence. *American Journal of Theology & Philosophy* 27(2–3): 259–284.

Cooey, Paula M., Sharon A. Farmer, and Mary Ellen Ross, eds. 1987. *Embodied love: Sensuality and relationship as feminist values*. San Francisco: Harper and Row.

Cooper-White, Pamela. 2007. *Many voices: Pastoral psychotherapy in relational and theological perspective.* Minneapolis, MN: Fortress.

Craig, William Lane. 2001a. *Time and eternity: Exploring God's relationship to time.* Wheaton, IL: Crossway Books.

———, ed. 2001b. *Time and the metaphysics of relativity.* Dordrecht: Kluwer.

Cramer, John G. 1986. The transactional interpretation of quantum mechanics. *Reviews of Modern Physics* 58: 647–687.

Cunningham, David S. 1998. *These three are one: The practice of trinitarian theology.* Oxford: Blackwell.

Cushing, James T. 1989. A background essay. In *Philosophical consequences of quantum theory: Reflections on Bell's theorem,* ed. J.T. Cushing and E. McMullin. Notre Dame, IN: University of Notre Dame Press.

———. 1994. *Quantum Mechanics: Historical contingency and the Copenhagen hegemony.* Chicago: University of Chicago Press.

———. 2001. Determinism versus indeterminism in quantum mechanics: A "free" choice. In *Quantum mechanics: Scientific perspectives on divine action,* ed. R.J. Russell, P. Clayton, K. Wegter-McNelly, and J. Polkinghorne. Vatican City State: Vatican Observatory/Center for Theology and the Natural Sciences.

Davies, Paul C. 1992. *The mind of God: The scientific basis for a rational world.* New York: Simon and Schuster.

———. 2009. The quantum life. *PhysicsWorld* 22 (7): 24–29.

Deane-Drummond, Celia. 2008. *Eco-theology.* London: Darton Longman and Todd.

Descartes, René. 1993. *Meditations on first philosophy.* 3rd ed. Trans. D.A. Cress. Indianapolis: Hackett.

d'Espagnat, Bernard. 2006. *On physics and philosophy.* Princeton, NJ: Princeton University Press.

DeWitt, Bryce S., Neill Graham, and Hugh Everett. 1973. *The many-worlds interpretation of quantum mechanics.* Princeton, NJ: Princeton University Press.

DeWitt, Calvin B. 2007. *Earth-wise: A biblical response to environmental issues.* 2nd ed. Grand Rapids, MI: Faith Alive Christian Resources.

Dickson, William Michael. 1998. *Quantum chance and non-locality: Probability and non-locality in the interpretations of quantum mechanics.* Cambridge: Cambridge University Press.

Dirac, Paul A.M. 1958. *The principles of quantum mechanics.* 4th ed. Oxford: Clarendon.

Edwards, Denis. 1999. *The God of evolution: A trinitarian theology.* New York: Paulist Press.

———. 2004. *Breath of life: A theology of the Creator Spirit.* Maryknoll, NY: Orbis.

———. 2006. *Ecology at the heart of faith.* Maryknoll, NY: Orbis.

Egan, Greg. 2002. Schild's ladder/decoherence (technical notes). http://www.gregegan.net/SCHILD/Decoherence/DecoherenceNotes.html (accessed August 2, 2010).

Einstein, Albert. 1948. Quantenmechanik und Wirklichkeit. *Dialectica* 2: 320–324.

———. 1970. Remarks concerning the essays brought together in this co-operative volume. In *Albert Einstein: Philosopher-Scientist,* ed. P.A. Schilpp. LaSalle, IL: Open Court.

Einstein, Albert, Boris Podolsky, and Nathan Rosen. 1935. Can quantum-mechanical description of physical reality be considered complete? *Physical Review* 47: 777–780.

Ellis, George F.R., ed. 2002. *The far-future universe: Eschatology from a cosmic perspective.* Philadelphia: Templeton Foundation.

Ernst, Joy, and Eldon Ernst. 1982. Freeing one another, power in partnership. *Journal of Women and Religion* 2 (1): 29–36.

Esfeld, Michael. 2001. *Holism in philosophy of mind and philosophy of physics.* Dordrecht: Kluwer.

Faraci, Giuseppe, Diego Gutkowski, Salvatore Notarrigo, and Agata R. Pennisi. 1974. Angular-correlation of scattered annihilation photons, to test possibility of hidden variables in quantum-theory. *Applied Physics* 5 (1): 63–65.

Feynman, Richard P., Robert B. Leighton, and Matthew Sands. 1965. *Quantum mechanics.* Vol. III of *The Feynman lectures on physics.* Reading, MA: Addison-Wesley.

Fiddes, Paul S. 1992. *The creative suffering of God.* Oxford: Oxford University Press.

Fine, Arthur. 1982. Some local models for correlation experiments. *Synthese* 50: 279–294.

———. 1989a. Correlations and efficiency: Testing Bell inequalities. *Foundations of Physics* 19: 453–478.

———. 1989b. Do correlations need to be explained? In *Philosophical consequences of quantum theory: Reflections on Bell's theorem,* ed. J.T. Cushing and E. McMullin. Notre Dame, IN: University of Notre Dame Press.

———. 1996. *The shaky game: Einstein, realism, and the quantum theory.* 2nd ed. Chicago: University of Chicago Press.

Finkelstein, David. 2003. What is a photon? *Optics and Photonics News* 14 (10): 12–17.

Folse, Henry J. 1989. Bohr on Bell. In *Philosophical consequences of quantum theory: Reflections on Bell's theorem,* ed. J.T. Cushing and E. McMullin. Notre Dame, IN: University of Notre Dame Press.

Fontinell, Eugene. 1979. Towards an ethics of relationships. In *Situationism and the new morality,* ed. R. Cunningham. New York: Appleton-Century-Crofts.

Fowler, Dean. 1979. Process theology of interdependence. *Theological Studies* 40 (1): 44–58.

Fox, Patricia. 2001. *God as communion: John Zizioulas, Elizabeth Johnson, and the retrieval of the symbol of the triune God.* Collegeville, MN: Liturgical Press.

Freedman, Stuart J., and John F. Clauser. 1972. Experimental test of local hidden-variable theories. *Physical Review Letters* 28: 938–941.

Fretheim, Terence E. 2005. *God and world in the Old Testament: A relational theology of creation.* Nashville, TN: Abingdon.

Friedman, Thomas L. 2006. *The world is flat: A brief history of the twenty-first century.* Updated and exp. ed. New York: Farrar Straus and Giroux.

García-Patrón, R., J. Fiurášek, N.J. Cerf, J. Wenger, R. Tualle-Brouri, and Ph. Grangier. 2004. Proposal for a loophole-free Bell test using homodyne detection. *Physical Review Letters* 93 (13): 130–409.

Ghirardi, GianCarlo, Alberto Rimini, and Tullio Weber. 1986. Unified dynamics for microscopic and macroscopic systems. *Physical Review D* 34: 470–491.

Ghosh, R., and Leonard Mandel. 1987. Observation of nonclassical effects in the interference of two photons. *Physical Review Letters* 59: 1903–1905.

Giberson, Karl. 2008. *Saving Darwin: How to be a Christian and believe in evolution.* New York: HarperCollins.

Gilder, Louisa. 2008. *The age of entanglement: When quantum physics was reborn.* New York: Alfred A. Knopf.

Glauber, Roy J. 2007. One hundred years of light quanta. *Annalen der Physik* 16 (1): 6–24.

Godzieba, Anthony. 2003. Bodies and persons, resurrected and postmodern: Towards a relational eschatology. In *Theology and conversation: Towards a relational theology*, ed. J. Haers and P. de Mey. Leuven: Leuven University/Peeters.

Gómez, Javier, Antonio Peimbert, and Juan Echevarría. 2009. Optical quantum entanglement in astrophysics. *Revista Mexicana de Fisica y Astrofísica* 45 (2): 179–189.

Greenberger, Daniel M., Michael A. Horne, Abner Shimony, and Anton Zeilinger. 1990. Bell's theorem without inequalities. *American Journal of Physics* 58 (12): 1131–1143.

Greenstein, George, and Arthur G. Zajonc. 1997. *The quantum challenge: Modern research on the foundations of quantum mechanics.* Boston: Jones and Bartlett.

Grenz, Stanley J. 1999. "Scientific" theology/"Theological" science: Pannenberg and the dialogue between theology and science. *Zygon* 34 (1): 159–166.

———. 2000. *Theology for the community of God.* Grand Rapids, MI: Eerdmans.

———. 2001. *The social God and the relational self: A trinitarian theology of the imago Dei.* Louisville, KY: Westminster/John Knox.

Grey, Mary C. 1991. Claiming power-in-relation: Exploring the ethics of connection. *Journal of Feminist Studies in Religion* 7 (1): 7–18.

———. 1999. Expelled again from Eden: Facing difference through connection. *Feminist Theology* 21: 8–20.

Grib, Andrei A., and Waldyr A. Rodrigues. 1999. *Nonlocality in quantum physics.* Dordrecht: Kluwer.

Grigg, Richard. 1994. Enacting the divine: Feminist theology and the being of God. *Journal of Religion* 74 (4): 506–523.

Gunton, Colin E. 1993. *The one, the three, and the many: God, creation, and the culture of modernity.* Cambridge: Cambridge University Press.

———. 1998. *The triune creator: A historical and systematic study.* Grand Rapids, MI: Eerdmans.

———. 2003. *The promise of trinitarian theology.* 2nd ed. Edinburgh: T & T Clark.

Haers, Jacques, and Peter de Mey, eds. 2003. *Theology and conversation: Towards a relational theology.* Leuven: Leuven University/Peeters.

Hall, Douglas John. 1986. *Imaging God: Dominion as stewardship.* New York: Eerdmans.

Hallman, David G. 1994. *Ecotheology: Voices from South and North.* Geneva: WCC Publications.

Hameroff, Stuart, and Roger Penrose. 1996. Orchestrated reduction of quantum coherence in brain microtubules: A model for consciousness. *Mathematics and Computers in Simulation* 40 (3–4): 453–480.

Hampson, Daphne. 1989. The theological implications of a feminist ethic. *Modern Churchman* 31 (1): 36–39.

———. 1993. Theological integrity and human relationships. *Feminist Theology* 2: 42–56.

Happel, Stephen. 1995. Divine providence and instrumentality: Metaphors for time in self-organizing systems and divine action. In *Chaos and complexity: Scientific perspectives on divine action*, ed. R.J. Russell, N. Murphy, and A. Peacocke. Vatican City State: Vatican Observatory/Center for Theology and the Natural Sciences.

———. 1996. Metaphors and time asymmetry: Cosmologies in physics and Christian meanings. In *Quantum cosmology and the laws of nature: Scientific perspectives on divine action*, ed. R.J. Russell, N. Murphy, and C.J. Isham. Vatican City State: Vatican Observatory/Center for Theology and the Natural Sciences.

Hardy, Lucien. 1993. Nonlocality for two particles without inequalities for almost all entangled states. *Physical Review Letters* 71 (11): 1665–1668.

Hart, John. 2006. *Sacramental commons: Christian ecological ethics.* Lanham, MD: Rowman and Littlefield.

Hartshorne, Charles. 1948. *The divine relativity: A social conception of God.* New Haven, CT: Yale University Press.

Hasegawa, Yuji, Rudolf Loidl, Gerald Badurek, Matthias Baron, and Helmut Rauch. 2003. Violation of a Bell-like inequality in single-neutron interferometry. *Nature* 425 (6953): 45–48.

Hauerwas, Stanley. 2000. *A better hope: Resources for a church confronting capitalism, democracy, and postmodernity.* Grand Rapids, MI: Brazos.

Haught, John F. 2000. *God after Darwin: A theology of evolution.* Boulder, CO: Westview.

Healey, Richard A. 1989. *The philosophy of quantum mechanics: An interactive interpretation.* Cambridge: Cambridge University Press.

Hefner, Philip J. 1993. *The human factor: Evolution, culture, and religion.* Minneapolis, MN: Fortress.

Heim, S. Mark. 2001. *The depth of the riches: A trinitarian theology of religious ends.* Grand Rapids, MI: Eerdmans.

Heisenberg, Werner. 1958. *Physics and philosophy: The revolution in modern science.* New York: Harper.

Heller, Michael. 1996. *The new physics and a new theology.* Trans. G.V. Coyne, S.J., S. Giovannini, and T.M. Sierotowicz. Vatican City State: Vatican Observatory.

Herbert, Nick. 1985. *Quantum reality: Beyond the new physics.* Garden City, NY: Anchor/Doubleday.

Herrington, Elizabeth. 1996. Just good friends: Towards a lesbian and gay theology of relationships. *Feminist Theology* 11: 119–120.

Hesse, Mary B. 1988. Physics, philosophy, and myth. In *Physics, philosophy, and theology: A common quest for understanding,* ed. R.J. Russell, W.R. Stoeger, S.J., and G.V. Coyne, S.J. Vatican City State: Vatican Observatory.

Heyward, Carter. 1982. *The redemption of God: A theology of mutual relation.* Lanham, MD: University Press of America.

Hodgson, Peter E. 2005. *Theology and modern physics.* Aldershot, UK: Ashgate.

Holt, Richard A., and Francis M. Pipkin. 1974. Unpublished paper. Cambridge, MA.

Howard, Don. 1985. Einstein on locality and separability. *Studies in History and Philosophy of Science* 16: 171–201.

———. 1989. Holism, separability, and the metaphysical implications of the Bell experiments. In *Philosophical consequences of quantum theory: Reflections on Bell's theorem,* ed. J.T. Cushing and E. McMullin. Notre Dame, IN: University of Notre Dame Press.

———. 1990. *Nicht sein kann was nich sein darf,* or the prehistory of EPR, 1909–1935: Einstein's early worries about the quantum mechanics of composite systems. In *Sixty-two years of uncertainty: Historical, philosophical, and physical inquiries into the foundations of quantum mechanics,* ed. A.I. Miller. New York: Plenum.

———. 1993. Locality, separability, and the physical implications of the Bell experiments: a new interpretation. In *Bell's theorem and the foundations of modern physics: Proceedings of the 1991 Conference, 7–10 October, Cesena, Italy,* ed. A. van der Merwe, F. Selleri, and G. Tarrozi. Singapore: World Scientific.

Howell, Nancy R. 1989. Radical relatedness and feminist separatism. *Process Studies* 18 (2): 118–126.

———. 2000. *A feminist cosmology: Ecology, solidarity, and metaphysics.* Amherst, NY: Humanity/Prometheus.

Hughes, R.I.G. 1989a. Bell's theorem, ideology, and structural explanation. In *Philosophical consequences of quantum theory: Reflections on Bell's theorem,* ed. J.T. Cushing and E. McMullin. Notre Dame, IN: University of Notre Dame Press.

————. 1989b. *The structure and interpretation of quantum mechanics*. Cambridge, MA: Harvard University Press.

Hunt, Mary E. 1989. *Fierce tenderness: A feminist theology of friendship*. San Francisco: HarperCollins.

Hurley, Patrick J. 1979. Russell, Poincare, and Whitehead's relational theory of space. *Process Studies* 9 (1–2): 14–21.

Irenaeus. 1992. *Against the heresies*. Trans. D.J. Unger and J.J. Dillon. Vol. 55 of *Ancient Christian writers series*. New York: Paulist Press.

Jakobsen, Janet R. 1997. The gendered division of moral labor: Radical relationalism and feminist ethics. In *Living responsibly in community*, ed. F.E. Glennon and D. Trimiew. Lanham, MD: University Press of America.

Jammer, Max. 1974. *The philosophy of quantum mechanics: The interpretations of quantum mechanics in historical perspective*. Chichester, NY: Wiley.

Jansen, Henry. 1995. *Relationality and the concept of God*. Amsterdam: Rodopi.

Jarrett, Jon P. 1984. On the physical significance of the locality conditions in the Bell arguments. *Noûs* 18: 569–589.

————. 1989. Bell's theorem: A guide to the implications. In *Philosophical consequences of quantum theory: Reflections on Bell's theorem*, ed. J.T. Cushing and E. McMullin. Notre Dame, IN: University of Notre Dame Press.

Jeffries, Paul Christopher. 2000. The philosophical implications of quantum nonlocality. PhD diss., Cornell University.

Jenson, Robert W. 1982. *The triune identity: God according to the gospel*. Philadelphia: Fortress.

Johnson, Elizabeth A. 1992. *She who is: The mystery of God in feminist theological discourse*. New York: Crossroad.

Jones, Martin. 2002. Interpretations of entanglement. Paper delivered at the April meeting of the American Physical Society, jointly sponsored with the High Energy Astrophysics Division (HEAD) of the American Astronomical Society, Albuquerque, New Mexico.

Jüngel, Eberhard. 2001. *God's being is in becoming: The trinitarian being of God in the theology of Karl Barth*. Trans. J. Webster. Edinburgh: T and T Clark.

Kaiser, Christopher B. 1996. Quantum complementarity and Christological dialectic. In *Religion and science: History, method, dialogue*, ed. W.M. Richardson and W.J. Wildman. New York: Routledge.

Keller, Catherine. 1985. Wholeness and all the king's men. *Anima* 11 (2): 83–95.

————. 1987a. *From a broken web: Separation, sexism, and self*. Boston: Beacon.

————. 1987b. Walls, women and intimations of interconnection. In *Women in the world's religions past and present*. Ed. U. King. New York: Paragon House.

————. 1989. Feminism and the ethic of inseparability. In *Weaving the visions: New patterns in feminist spirituality*, ed. J. Plaskow and C. Christ. San Francisco: Harper and Row.

————. 1997. Composting our connections: Toward a spirituality of relation. In *Greening of faith*, ed. J.E. Carroll, P. Brockelman, M. Westfall, and B. McKibben. Hanover, NH: University Press of New England.

————. 2003. *Face of the deep: A theology of becoming*. New York: Routledge.

Keller, Catherine, and Laurel C. Schneider, eds. 2010. *Polydoxy: Theology of multiplicity and relation*. London: Routledge.

Korsmeyer, Jerry D. 1998. *Evolution & Eden: Balancing original sin and contemporary science*. New York: Paulist Press.

Krobath, Evi. 1988. Im Anfang ist die Beziehung. *Evangelische Theologie* 48 (1): 80–82.

LaCugna, Catherine Mowry. 1991. *God for us: The Trinity and Christian life*. San Francisco: HarperSanFrancisco.

Lamehi-Rachti, Mohammad, and Wolfgang Mittig. 1976. Quantum mechanics and hidden variables: A test of Bell's inequality by the measurement of the spin correlation in low-energy proton-proton scattering. *Physical Review D* 14: 2543–2555.

Leopold, Aldo. 1987. *A Sand County almanac and sketches here and there*. Oxford: Oxford University.

Loder, James Edwin, and Jim W. Neidhardt. 1996. Barth, Bohr, and dialectic. In *Religion and science: History, method, dialogue*, ed. W.M. Richardson and W.J. Wildman. New York: Routledge.

Loomer, Bernard. 1984. A process-relational conception of creation. In *Cry of the environment: Rebuilding the Christian creation tradition*, ed. P.N. Joranson and K. Butigan. Santa Fe, NM: Bear.

Lossky, Vladimir. 1957. *The mystical theology of the Eastern Church*. London: James Clarke.

Lowe, Mary E. 2000. Woman oriented hamartiologies: A survey of the shift from powerlessness to right relationship. *Dialog* 39 (2): 119–139.

Mackinnon, Edward. 1996. Complementarity. In *Religion and science: History, method, dialogue*, ed. W.M. Richardson and W.J. Wildman. New York: Routledge.

Maudlin, Tim. 2002. *Quantum non-locality and relativity: Metaphysical intimations of modern physics*. 2nd ed. Oxford: Blackwell.

McFague, Sallie. 1982. *Metaphorical theology: Models of God in religious language*. Philadelphia: Fortress.

———. 1987. *Models of God: Theology for an ecological, nuclear age*. Philadelphia: Fortress.

———. 1993. *The body of God: An ecological theology*. Minneapolis, MN: Fortress.

———. 1997. *Super, natural Christians: How we should love nature*. Minneapolis, MN: Fortress.

———. 2000. *Life abundant: Rethinking theology and economy for a planet in peril*. Minneapolis, MN: Fortress.

McGrath, Alister E. 2001. *Nature*. Vol. 1 of *A scientific theology*. Grand Rapids, MI: Eerdmans.

McMullin, Ernan. 1981. How should cosmology relate to theology? In *The sciences and theology in the twentieth century*, ed. A. Peacocke. Notre Dame, IN: University of Notre Dame Press.

———. 1989. The explanation of distant action: Historical notes. In *Philosophical consequences of quantum theory: Reflections on Bell's theorem*, ed. J.T. Cushing and E. McMullin. Notre Dame, IN: University of Notre Dame Press.

Mermin, N. David. 1985. Is the moon there when nobody looks? Reality and the quantum theory. *Physics Today* 38 (4): 38–47.

———. 1998. The Ithaca interpretation of quantum mechanics. *Pramana–Journal of Physics* 51 (5): 549–565.

———. 1999. What do these correlations know about reality, nonlocality and the absurd? *Foundations of Physics* 29 (4): 571–587.

Milbank, John. 2005. *Theology and social theory: Beyond secular reason*. 2nd ed. Oxford: Blackwell.

Milburn, Gerard J. 1996. *Schrödinger's machines: The quantum technology reshaping everyday life*. New York: W.H. Freeman.

Miller, Kenneth R. 1999. *Finding Darwin's God: A scientist's search for common ground between God and evolution*. New York: HarperCollins.

Moltmann, Jürgen. 1993a. *The crucified God: The cross of Christ as the foundation and criticism of Christian theology*. Trans. R.A. Wilson and J. Bowden. Minneapolis, MN: Fortress.

————. 1993b. *God in creation: A new theology of creation and the Spirit of God.* Trans. M. Kohl. Minneapolis, MN: Fortress.

————. 1993c. *Theology of hope: On the ground and the implications of a Christian eschatology.* Trans. J.W. Leitch. Minneapolis, MN: Fortress.

————. 1993d. *The Trinity and the kingdom: The doctrine of God.* Trans. M. Kohl. Minneapolis, MN: Fortress.

Murphy, George L., LaVonne Althouse, and Russell E. Willis. 1996. *Cosmic witness: Commentaries on science/technology themes.* Lima, OH: CSS.

Musgrave, Alan. 2006–2007. The "miracle argument" for scientific realism. *Rutherford Journal* 2. http://www.rutherfordjournal.org/article020108.html (accessed July 28, 2010).

Muthukrishnan, Ashok, Marlan O. Scully, and M. Suhail Zubairy. 2003. The concept of the photon—revisited. *Optics and Photonics News* 14 (10): 18–27.

Neuger, Christie Cozad. 1999. Women and relationality. In *Feminist and womanist pastoral theology*, ed. B. Gill-Austern and B. Miller-McLemore. Nashville, TN: Abingdon.

Nicolaidis, Argyris. 2010. Relational nature. In *The Trinity and an entangled world: Relationality in physical science and theology*, ed. J. Polkinghorne. Grand Rapids, MI: Eerdmans.

Niebuhr, Reinhold. 1964. *Human destiny.* Vol. 2 of *The nature and destiny of man: A Christian interpretation.* New York: Scribner's Sons.

Norris, Christopher. 2000. *Quantum theory and the flight from realism: Philosophical responses to quantum mechanics.* London: Routledge.

Norris, Thomas J. 2009. *The Trinity, life of God, hope for humanity: Towards a theology of communion.* Hyde Park, NY: New City.

O'Connell, A.D., M. Hofheinz, M. Ansmann, Radoslaw C. Bialczak, M. Lenander, Erik Lucero, M. Neeley, D. Sank, H. Wang, M. Weides, J. Wenner, John M. Martinis, and A.N. Cleland. 2010. Quantum ground state and single-phonon control of a mechanical resonator. *Nature* 464 (7289): 697–703.

Oliver, Harold H. 1981. *A relational metaphysic.* The Hague: M. Nijhoff.

————. 1984. *Relatedness: Essays in metaphysics and theology.* Macon, GA: Mercer University.

————. 2006. *Metaphysics, theology, and self: Relational essays.* Macon, GA: Mercer University.

Oord, Thomas Jay, and Michael E. Lodahl. 2005. *Relational holiness: Responding to the call of love.* Kansas City, MO: Beacon Hill.

Pannenberg, Wolfhart. 1977. *Jesus—God and man.* 2nd ed. Trans. L. Wilkins and D. Priebe. Philadelphia: Westminster.

————. 1988. A response to my American friends. In *The theology of Wolfhart Pannenberg: Twelve American critiques, with an autobiographical essay and response*, ed. C.E. Braaten and P. Clayton. Minneapolis, MN: Augsburg.

————. 1991. *Systematic theology.* Vol. 1. Trans. G.W. Bromiley. Grand Rapids, MI: Eerdmans.

————. 1993. *Toward a theology of nature: Essays on science and faith.* Ed. T. Peters. Louisville, KY: Westminster/John Knox.

————. 1994. *Systematic theology.* Vol. 2. Trans. G.W. Bromiley. Grand Rapids, MI: Eerdmans.

Peacocke, Arthur. 1984. *Intimations of reality: Critical realism in science and religion.* Notre Dame, IN: University of Notre Dame Press.

————. 1993. *Theology for a scientific age: Being and becoming—natural, divine and human.* Minneapolis, MN: Fortress.

Penrose, Roger. 2005. *The road to reality: A complete guide to the laws of the universe.* New York: Alfred A. Knopf.

Peters, Ted. 1988. McFague's metaphors. *Dialog* 27 (2): 130–140.

———. 1993. *God as Trinity: Relationality and temporality in the divine life.* Louisville, KY: Westminster/John Knox.

———. 1997. Clarity of the part versus meaning of the whole. In *Beginning with the end: God, science, and Wolfhart Pannenberg,* ed. C.R. Albright and J. Haugen. Chicago: Open Court.

———. 2000. *God—the world's future: Systematic theology for a postmodern era.* 2nd ed. Minneapolis, MN: Fortress.

———. 2002. *Playing God? Genetic determinism and human freedom.* 2nd ed. New York: Routledge.

Peters, Ted, and Martinez J. Hewlett. 2003. *Evolution from creation to new creation: Conflict, conversation, and convergence.* Nashville, TN: Abingdon.

Peters, Ted, Robert J. Russell, and Michael Welker, eds. 2002. *Resurrection: Theological and scientific assessments.* Grand Rapids, MI: Eerdmans.

Pinnock, Clark H. 2001. *Most moved mover: A theology of God's openness.* Grand Rapids, MI: Baker Academic.

Placher, William C. 1994. *Narratives of a vulnerable God: Christ, theology, and scripture.* Louisville, KY: Westminster/John Knox.

Polkinghorne, John. 1984. *The quantum world.* Princeton, NJ: Princeton University Press.

———. 1989. *Science and providence: God's interaction with the world.* Boston: Shambhala Publications.

———. 2000. *Faith, science, and understanding.* New Haven, CT: Yale University Press.

———. 2001a. Kenotic creation and divine action. In *The work of love: Creation as kenosis,* ed. J. Polkinghorne. Grand Rapids, MI: Eerdmans.

———, ed. 2001b. *The work of love: Creation as kenosis.* Grand Rapids, MI: Eerdmans.

———. 2002. *Quantum theory: A very short introduction.* Oxford: Oxford University.

———. 2005. *Exploring reality: The intertwining of science and religion.* New Haven, CT: Yale University Press.

———. 2007. *Quantum physics and theology: An unexpected kinship.* New Haven, CT: Yale University Press.

———. 2010a. The demise of Democritus. In *The Trinity and an entangled world: Relationality in physical science and theology,* ed. J. Polkinghorne. Grand Rapids, MI: Eerdmans.

———, ed. 2010b. *The Trinity and an entangled world: Relationality in physical science and theology.* Grand Rapids, MI: Eerdmans.

Polkinghorne, John, and Michael Welker, eds. 2000. *The end of the world and the ends of God: Science and theology on eschatology.* Harrisburg, PA: Trinity.

Pratt, Douglas. 2002. *Relational deity: Hartshorne and Macquarrie on God.* Lanham, MD: University Press of America.

Primack, Joel, and Nancy Abrams. 2006. *The view from the center of the universe: Discovering our extraordinary place in the cosmos.* New York: Riverhead Books.

Prokes, Mary Timothy. 1993. *Mutuality: The human image of trinitarian love.* New York: Paulist Press.

Putnam, Hilary. 1975. *Mathematics, matter and method: Philosophical papers.* Vol. 1. Cambridge: Cambridge University Press.

Rahner, Karl. 1970. *The Trinity.* Trans. J. Donceel. London: Burns and Oates.

———. 1976. *Foundations of Christian faith: An introduction to the idea of Christianity.* Trans. W.V. Dych. New York: Crossroad.

Raitt, Jill. 1982. Strictures and structures: Relational theology and a woman's contribution to theological conversation. *Journal of the American Academy of Religion* 50 (1): 3–17.

Rasmussen, Larry L. 1996. *Earth community earth ethics*. Maryknoll, NY: Orbis.

Rauch, Helmut, Anton Zeilinger, Gerald Badurek, A. Wilding, W. Bauspiess, and U. Bonse. 1975. Verification of coherent spinor rotation of fermions. *Physics Letters A* 54 (6): 425–427.

Redhead, Michael L.G. 1987. *Incompleteness, nonlocality, and realism: A prolegomenon to the philosophy of quantum mechanics*. Oxford: Clarendon.

———. 2001. The tangled story of nonlocality in quantum mechanics. In *Quantum mechanics: Scientific perspectives on divine action*, ed. R.J. Russell, P. Clayton, K. Wegter-McNelly, and J. Polkinghorne. Vatican City State: Vatican Observatory/Center for Theology and the Natural Sciences.

Reeves, Gene. 1986. To be is to be for others. *American Journal of Theology and Philosophy* 7 (1): 41–45.

Ricoeur, Paul. 1975. Biblical hermeneutics. *Semeia* 4: 29–148.

Rosenblum, Bruce, and Fred Kuttner. 2006. *Quantum enigma: Physics encounters consciousness*. Oxford: Oxford University Press.

Rosenfeld, Léon. 1983. The Einstein–Podolsky–Rosen paper. In *Quantum theory and measurement*, ed. J.A. Wheeler and W.H. Zurek. Princeton, NJ: Princeton University Press.

Rowe, Mary A., David Kielpinski, Volker Meyer, Cass Sackett, Wayne M. Itano, Christopher Monroe, and David Wineland. 2001. Experimental violation of Bell's inequality with efficient detection. *Nature* 409: 791–794.

Roxburgh, Alan. 2000. Rethinking trinitarian missiology. In *Global missiology for the 21st century: The Iguassu dialogue*, ed. W.D. Taylor. Grand Rapids, MI: Baker Academic.

Ruether, Rosemary Radford. 1992. *Gaia & God: An ecofeminist theology of earth healing*. San Francisco: HarperCollins.

Russell, Letty M. 1979. *The future of partnership*. Philadelphia: Westminster.

———. 1981. *Growth in partnership*. Philadelphia: Westminster.

Russell, Robert J. 1985. The physics of David Bohm and its relevance to philosophy and theology. *Zygon* 20 (2): 135–158.

———. 1988. Quantum physics in philosophical and theological perspective. In *Physics, philosophy, and theology: A common quest for understanding*, ed. R.J. Russell, W.R. Stoeger, S.J., and G.V. Coyne, S.J. Vatican City State: Vatican Observatory.

———. 1989. Cosmology, creation, and contingency. In *Cosmos as creation: Theology and science in consonance*, ed. T. Peters. Nashville, TN: Abingdon.

———. 2001. Divine action and quantum mechanics: A fresh assessment. In *Quantum mechanics: Scientific perspectives on divine action*, ed. R.J. Russell, P. Clayton, K. Wegter-McNelly, and J. Polkinghorne. Vatican City State: Vatican Observatory/Center for Theology and the Natural Sciences.

———. 2002. Bodily resurrection, eschatology, and scientific cosmology. In *Resurrection: Theological and scientific assessments*, ed. T. Peters, R.J. Russell, and M. Welker. Grand Rapids, MI: Eerdmans.

———. 2006. *Cosmology, evolution, and resurrection hope: Theology and science in creative mutual interaction*. Kitchener, ON: Pandora.

———. Forthcoming. *Time in eternity*. Notre Dame, IN: University of Notre Dame Press.

Russell, Robert J., Philip Clayton, Kirk Wegter-McNelly, and John Polkinghorne, eds. 2001. *Quantum mechanics: Scientific perspectives on divine action*. Vatican City State: Vatican Observatory/Center for Theology and the Natural Sciences.

Russell, Robert J., Nancey Murphy, and Chris J. Isham, eds. 1996. *Quantum cosmology and the laws of nature: Scientific perspectives on divine action*. 2nd ed.

Vatican City State: Vatican Observatory/Center for Theology and the Natural Sciences.

Russell, Robert J., Nancey Murphy, Theo C. Meyering, and Michael A. Arbib, eds. 1999. *Neuroscience and the person: Scientific perspectives on divine action.* Vatican City State: Vatican Observatory/Center for Theology and the Natural Sciences.

Russell, Robert J., Nancey Murphy, and Arthur Peacocke, eds. 1995. *Chaos and complexity: Scientific perspectives on divine action.* Vatican City State: Vatican Observatory/Center for Theology and the Natural Sciences.

Russell, Robert J., Nancey Murphy, and William R. Stoeger, eds. 2008. *Scientific perspectives on divine action: Twenty years of challenge and progress.* Vatican City State: Vatican Observatory/Center for Theology and the Natural Sciences.

Russell, Robert J., William R. Stoeger, S.J., and Francisco J. Ayala, eds. 1998. *Evolutionary and molecular biology: Scientific perspectives on divine action.* Vatican City State: Vatican Observatory/Center for Theology and the Natural Sciences.

Russell, Robert J., William R. Stoeger, S.J., and George V. Coyne, S.J., eds. 1988. *Physics, philosophy, and theology: A common quest for understanding.* Vatican City State: Vatican Observatory.

Salart, Daniel, Augustin Baas, Cyril Branciard, Nicolas Gisin, and Hugo Zbinden. 2008. Testing the speed of "spooky action at a distance." *Nature* 454 (14): 861–864.

Sanders, John. 1998. *The God who risks: A theology of divine providence.* Downers Grove, IL: IVP Academic.

Sarovar, Mohan, Akihito Ishizaki, Graham R. Fleming, and K. Birgitta Whaley. 2010. Quantum entanglement in photosynthetic light-harvesting complexes. *Nature Physics* 6 (6): 462–467.

Sawtelle, Roger A. 1997. *The God who relates: An African-American trinitarian theology.* Lowell, MA: Samizdat.

Scarani, Valerio. 2006. *Quantum physics: A first encounter: Interference, entanglement, and reality.* Trans. R. Thew. Oxford: Oxford University Press.

Schmitz-Moormann, Karl. 1997. *Theology of creation in an evolutionary world.* Cleveland, OH: Pilgrim.

Schrödinger, Erwin. 1935. Discussion of probability relations between separated systems. *Proceedings of the Cambridge Philosophical Society* 31: 555–563.

———. 1983. The present situation in quantum mechanics. In *Quantum theory and measurement,* ed. J.A. Wheeler and W.H. Zurek. Trans. J.D. Trimmer. Princeton, NJ: Princeton University Press.

Scully, Marlan O., and Kai Druhl. 1982. Quantum eraser: A proposed photon correlation experiment concerning observation and "delayed choice" in quantum mechanics. *Physical Review A* 25: 2208–2213.

Sharpe, Kevin. 1993. *David Bohm's world: New physics and new religion.* Lewisburg, PA: Bucknell University.

Sharpe, Kevin, and Jon Walgate. 1999. Patterns of the real: Quantum nonlocality. *Science & Spirit* 10 (1): 10–12.

Shimony, Abner. 1989. Search for a worldview which can accommodate our knowledge of microphysics. In *Philosophical consequences of quantum theory: Reflections on Bell's theorem,* ed. J.T. Cushing and E. McMullin. Notre Dame, IN: University of Notre Dame Press.

———. 1993. *Search for a naturalistic world view.* 2 vols. Cambridge: Cambridge University Press.

———. 2001. The reality of the quantum world. In *Quantum mechanics: Scientific perspectives on divine action,* ed. R.J. Russell, P. Clayton, K. Wegter-McNelly, and J. Polkinghorne. Vatican City State: Vatican Observatory/Center for Theology and the Natural Sciences.

———. 2004. Bell's theorem (fall 2006 edition). *Stanford Encyclopedia of Philosophy.* http://plato.standford.edu/entries/bell-theorem/ (accessed June 20, 2008).

Shults, F. LeRon. 2003. *Reforming theological anthropology: After the philosophical turn to relationality.* Grand Rapids, MI: Eerdmans.

———. 2005. *Reforming the doctrine of God.* Grand Rapids, MI: Eerdmans.

———. 2008. *Christology and science.* Grand Rapids, MI: Eerdmans.

Shults, F. LeRon, and Steven J. Sandage. 2003. *The faces of forgiveness: Searching for wholeness and salvation.* Grand Rapids, MI: Baker Academic.

———. 2006. *Transforming spirituality: Integrating theology and psychology.* Grand Rapids, MI: Baker Academic.

Simmons, Ernest L. 1999. Toward a kenotic pneumatology: Quantum field theory and the theology of the cross. *CTNS Bulletin* 19 (2): 11–16.

———. 2000. The sighs of God: Kenosis, quantum field theory, and the spirit. *Word and World.* Suppl. 4: 182–191.

———. 2006. Quantum *perichoresis:* Quantum field theory and the Trinity. *Theology and Science* 4 (2): 137–150.

Simon, Christoph, and William T.M. Irvine. 2003. Robust long-distance entanglement and a loophole-free Bell test with ions and photons. *Physical Review Letters* 91 (11): 110–405.

Smith, Archie. 1982. *The relational self: Ethics & therapy from a Black church perspective.* Nashville, TN: Abingdon.

Sobosan, Jeffrey G. 1999. *Romancing the universe: Theology, science, and cosmology.* Grand Rapids, MI: Eerdmans.

Soskice, Janet M. 1988. Knowledge and experience in science and religion: Can we be realists? In *Physics, philosophy, and theology: A common quest for understanding,* ed. R.J. Russell, W.R. Stoeger, S.J., and G.V. Coyne, S.J. Vatican City State: Vatican Observatory.

Spalding, Anne. 1999a. Being part of "right relation." *Feminist Theology* 22 (Summer): 9–24.

———. 1999b. "Right relation" revisited: Implications of right relation in the practice of church and Christian perceptions of God. *Feminist Theology* 28 (Summer): 57–68.

Sponheim, Paul R. 1993. *Faith and the other: A relational theology.* Minneapolis, MN: Fortress.

———. 2006. *Speaking of God: Relational theology.* St. Louis, MO: Chalice.

Stapp, Henry P. 1977. Quantum mechanics, local causality, and process philosophy. *Process Studies* 7 (3): 173–182.

———. 1989. Quantum nonlocality and the description of nature. In *Philosophical consequences of quantum theory: Reflections on Bell's theorem,* ed. J.T. Cushing and E. McMullin. Notre Dame, IN: University of Notre Dame Press.

Sturm, Douglas. 1998. *Solidarity and suffering: Toward a politics of relationality.* Albany: State University of New York Press.

Suchocki, Marjorie. 1985. Weaving the world. *Process Studies* 14 (2): 76–86.

———. 1989. *God, Christ, church: A practical guide to process theology.* New rev. ed. New York: Crossroad.

———. 1994. *The fall to violence: Original sin in relational theology.* New York: Continuum.

Sullivan, John A. 1987. *Explorations in Christology: The impact of process/relational thought.* New York: Peter Lang.

Teilhard de Chardin, Pierre. 1975. *The phenomenon of man.* Trans. B. Wall. New York: Harper and Row.

Teller, Paul. 1989. Relativity, relational holism, and the Bell inequalities. In *Philosophical consequences of quantum theory: Reflections on Bell's theorem*, ed. J.T. Cushing and E. McMullin. Notre Dame, IN: University of Notre Dame Press.

Tillich, Paul. 1951. *Systematic theology*. Vol. 1. Chicago: University of Chicago Press.

———. 1963. *Systematic theology*. Vol. 3. Chicago: University of Chicago Press.

Tittel, Wolfgang, Jürgen Brendel, Bernard Gisin, Thomas Herzog, Hugo Zbinden, and Nicolas Gisin. 1998. Experimental demonstration of quantum correlations over more than 10 km. *Physical Review A* 57 (5): 3229–3232.

Torrance, Alan J. 1996. *Persons in communion: An essay on trinitarian description and human participation with special reference to volume one of Karl Barth's Church dogmatics*. Edinburgh: T & T Clark.

Tracy, David. 1981. *The analogical imagination: Christian theology and the culture of pluralism*. New York: Crossroad.

Ursin, R., F. Tiefenbacher, T. Schmitt-Manderbach, H. Weier, T. Scheidl, M. Lindenthal, B. Blauensteiner, T. Jennewein, J. Perdigues, P. Trojek, B. Omer, M. Furst, M. Meyenburg, J. Rarity, Z. Sodnik, C. Barbieri, H. Weinfurter, and A. Zeilinger. 2007. Entanglement-based quantum communication over 144 km. *Nature Physics* 3 (7): 481–486.

van Fraassen, Bas C. 1980. *The scientific image*. Oxford: Oxford University Press.

———. 1989. The Charybdis of realism: Epistemological implications of Bell's inequality. In *Philosophical consequences of quantum theory: Reflections on Bell's theorem*, ed. J.T. Cushing and E. McMullin. Notre Dame, IN: University of Notre Dame Press.

Van Huyssteen, J. Wentzel. 1989. *Theology and the justification of faith: Constructing theories in systematic theology*. Grand Rapids, MI: Eerdmans.

———. 1999. *The shaping of rationality: Toward interdisciplinarity in theology and science*. Grand Rapids, MI: Eerdmans.

Van Raamsdonk, Mark. 2010. Building up spacetime with quantum entanglement. *General Relativity and Gravitation* 42 (10): 2323–2329.

Volf, Miroslav. 1996. *Exclusion and embrace: A theological exploration of identity, otherness, and reconciliation*. Nashville, TN: Abingdon.

———. 1998. *After our likeness: The church as the image of the Trinity*. Grand Rapids, MI: Eerdmans.

von Neumann, John. 1955. *Mathematical foundations of quantum mechanics*. Trans. R.T. Beyer. Princeton, NJ: Princeton University Press.

Voskuil, Duane. 2007. Entanglement, Slits and Buckyballs. *Process Studies* 36 (1): 23–44.

Walsh, Sylvia I. 1982. Women in love. *Soundings* 65 (3): 352–368.

Ware, Kallistos. 2010. The Holy Trinity: Model for personhood-in-relation. In *The Trinity and an entangled world: Relationality in physical science and theology*, ed. J. Polkinghorne. Grand Rapids, MI: Eerdmans.

Webb, Stephen H. 1996. *The gifting God: A trinitarian ethics of excess*. Oxford: Oxford University Press.

Wegter-McNelly, Kirk. 2000. Difference within theology nature: The strategies of intelligibility and credibility. *Journal of Faith and Science Exchange* 4: 241–264.

———. 2007. Natural evil in a divinely entangled world. In *Physics and cosmology: Scientific perspectives on the problem of natural evil*, ed. N. Murphy, R.J. Russell, and W.R. Stoeger. Vatican City State: Vatican Observatory/Center for Theology and the Natural Sciences.

———. 2008. Does God need room to act? Theo-physical in/compatibilism in non-interventionist theories of objectively special divine action. In *Scientific perspectives on divine action: Twenty years of challenge and progress*. ed. R.J. Russell, N. Murphy, and W.R. Stoeger. Vatican City State: Vatican Observatory/Center for Theology and the Natural Sciences.

Weihs, Gregor, Thomas Jennewein, Christoph Simon, Harald Weinfurter, and Anton Zeilinger. 1998. Violation of Bell's inequality under strict Einstein locality conditions. *Physical Review Letters* 81 (23): 5039–5043.

Weisskopf, Victor. 1992. There is a Bohr complementarity between science and religion. In *Cosmos, bios, theos: Scientists reflect on science, God, and the origins of the universe, life, and Homo sapiens*, ed. H. Margenau and R.A. Varghese. LaSalle, IL: Open Court.

Welch, Sharon D. 1990. *A feminist ethic of risk*. Minneapolis, MN: Fortress.

Welker, Michael. 2010. Relation: Human and divine. In *The Trinity and an entangled world: Relationality in physical science and theology*, ed. J. Polkinghorne. Grand Rapids, MI: Eerdmans.

Wheeler, David L. 1989. *A relational view of the atonement: Prolegomenon to a reconstruction of the doctrine*. New York: P. Lang.

Wheeler, John Archibald. 1946. Polyelectrons. *Annals of the New York Academy of Sciences* 48: 219.

Wheeler, John Archibald, and Wojciech Hubert Zurek, eds. 1983. *Quantum theory and measurement*. Princeton, NJ: Princeton University Press.

Whitaker, Andrew. 1998. John Bell and the most profound discovery of science. *PhysicsWorld* 11 (12): 29–34.

Whitehead, Alfred North. 1957. *Process and reality*. New York: Free Press.

———. 1967. *Science and the modern world*. New York: Free Press.

Wildman, Wesley J. 2010. An introduction to relational ontology. In *The Trinity and an entangled world: Relationality in physical science and theology*, ed. J. Polkinghorne. Grand Rapids, MI: Eerdmans.

Will, Clifford M. 2006. The confrontation between general relativity and experiment. *Living Reviews in Relativity* 9 (3). http://www.livingreviews.org/lrr-2006-3 (accessed August 24, 2010).

Will, James E. 1994. *The universal God: Justice, love, and peace in the global village*. Louisville, KY: Westminster/John Knox.

Wilson-Kastner, Patricia. 1983. *Faith, feminism, and the Christ*. Philadelphia: Fortress.

Winsberg, Eric, and Arthur Fine. 2003. Quantum life: Interaction, entanglement, and separation. *Journal of Philosophy* C (2): 80–97.

Wright, N.T. 2008. *Surprised by hope: Rethinking heaven, the resurrection, and the mission of the Church*. New York: HarperCollins.

Wu, Chien-Shiung, and Irving Shaknov. 1950. The angular correlation of scattered annihilation radiation. *Physical Review* 77: 136.

Yeager, Diane M. 1988. The web of relationship: Feminists and Christians. *Soundings* 71 (4): 485–513.

Zajonc, Arthur G. 2003. Light reconsidered. *Optics and Photonics News* 14 (10): 2–5.

Zbinden, Hugo, J. Brendel, Nicolas Gisin, and Wolfgang Tittel. 2001. Experimental test of nonlocal quantum correlation in relativistic configurations. *Physical Review A* 63 (2): 1–10.

Zeilinger, Anton. 1986. Testing Bell's inequalities with periodic switching. *Physics Letters A* 118: 1.

———. 2010. Quantum physics: Ontology or epistemology? In *The Trinity and an entangled world: Relationality in physical science and theology*, ed. J. Polkinghorne. Grand Rapids, MI: Eerdmans.

Zeilinger, Anton, Gregor Weihs, Thomas Jennewein, and Markus Aspelmeyer. 2005. Happy centenary, photon. *Nature* 433: 230–242.

Zizioulas, John D. 1985. *Being as communion: Studies in personhood and the church.* Crestwood, NY: St. Vladimir's Seminary.

———. 2006. *Communion and otherness.* New York: Continuum.

———. 2010. Relational ontology: Insights from patristic thought. In *The Trinity and an entangled world: Relationality in physical science and theology*, ed. J. Polkinghorne. Grand Rapids, MI: Eerdmans.

Zurek, Wojciech H. 2002. Decoherence and the transition from quantum to classical—revisited. *Los Alamos Science* 27: 1–25.

Zycinski, Joseph M. 1999. The laws of nature and the immanence of god in the evolving universe. In *The interplay between scientific and theological worldviews*, ed. N.H. Gregersen, U. Görman and C. Wassermann. Vol. 5 of *Studies in Science and Theology.* Geneva: Labor et Fides.

Index